The Green Revolution

Innovative Techniques for Sustainable Farming Practices

Magnus Massy

loss due to the information herein, either directly or indirectly. Respective authors own all copyrights not held by the publisher. The information herein is offered for informational purposes solely, and is universal as so. The presentation of the information is without contract or any type of guarantee assurance. The trademarks that are used are without any consent, and the publication of the trademark is without permission or backing by the trademark owner. All trademarks and brands within this book are for clarifying purposes only and are the owned by the owners themselves, not affiliated with this document.

Table of Contents

Chapter 1

Understanding the Green Revolution

Origins and Evolution

The roots of the Green Revolution stretch back to a period when global agriculture faced a formidable challenge: feeding a rapidly growing population. Post-World War II, the world was in a state of rebuilding, and the agricultural landscape was no exception. The Green Revolution, a term coined in the 1960s, heralded a significant transformation in agricultural practices. Driven by the need to increase food production, it marked the advent of new technologies and methodologies that revolutionized traditional farming.

The journey began in Mexico in the 1940s, where American agronomist Norman Borlaug spearheaded efforts to develop high-yielding varieties of wheat. Borlaug's work was instrumental in addressing the dire food shortages that plagued the region. Through cross-breeding techniques and the introduction of disease-resistant strains, he significantly boosted wheat production. This success laid the foundation for similar agricultural advancements in other parts of the world, particularly in Asia and Latin America.

At its core, the Green Revolution was fueled by the introduction of high-yield variety (HYV) seeds. These seeds were engineered to produce more grain per plant, thus increasing overall yield. Alongside these seeds, the widespread use of chemical fertilizers, pesticides, and irrigation systems became prevalent. These inputs were crucial in enhancing crop productivity and combating the limitations posed by traditional farming methods. However, the implementation of these technologies required significant investment and infrastructure, which was not always accessible to smallholder farmers.

The impact of the Green Revolution on global agriculture was profound. Countries like India and Pakistan, once on the brink of famine, witnessed dramatic increases in food production. In the 1960s and 1970s, these nations transformed from food-deficient to self-sufficient, and in some cases, even surplus-producing. This agricultural boom played a pivotal role in alleviating hunger and malnutrition, improving food security for millions of people.

Despite its successes, the Green Revolution was not without its critics. The heavy reliance on chemical inputs raised environmental concerns, as overuse led to soil degradation, water pollution, and a reduction in biodiversity. The monoculture practices promoted during this era often resulted in the loss of traditional crop varieties and a decrease in genetic diversity. Additionally, the benefits of the Green Revolution were not evenly distributed. Wealthier farmers with access to resources reaped the rewards, while poorer, small-scale farmers struggled to keep up with the costs associated with the new technologies.

The Green Revolution also brought about significant social changes. The increased mechanization of agriculture led to the displacement of rural laborers, as machines began to replace manual labor. This shift contributed to urban migration, with many former farmworkers seeking employment in cities. Furthermore, the focus on staple crops like wheat and rice, while boosting caloric intake, often overshadowed the cultivation of nutrient-rich traditional crops, affecting dietary diversity.

As the world grappled with these challenges, a new wave of innovation began to emerge. Researchers and policymakers recognized the need to address the environmental and social issues that accompanied the Green Revolution. This realization paved the way for the development of sustainable farming practices that aimed to balance productivity with ecological and social responsibility.

The evolution of the Green Revolution into a more sustainable model involved integrating traditional knowledge with modern scientific advancements. Efforts were made to promote agroecological practices, such as crop rotation and polyculture, which enhance soil fertility and reduce reliance on chemical inputs. The use of organic fertilizers and natural pest control methods gained traction, as they offered environmentally friendly alternatives to synthetic chemicals.

Water management also became a focal point in the evolution of sustainable agriculture. The Green Revolution's heavy reliance on irrigation had led to the depletion of water resources in many regions. Innovative techniques, such as drip irrigation and rainwater harvesting, were introduced to optimize water use and reduce wastage. These methods not only conserved water but also improved crop yields by ensuring plants received adequate moisture.

The role of technology in the evolution of sustainable farming practices cannot be overstated. Precision agriculture, which utilizes data and technology to optimize farming practices, emerged as a game-changer. By employing sensors, drones, and satellite imagery, farmers could monitor crop health, soil conditions, and weather patterns in real-time. This information allowed for targeted interventions, reducing the need for blanket applications of fertilizers and pesticides.

The Green Revolution's legacy also prompted a reevaluation of agricultural policies and investment strategies. Governments and international organizations recognized the importance of supporting smallholder farmers, who constitute a significant portion of the global agricultural workforce. Initiatives were launched to provide access to credit, training, and resources, empowering these farmers to adopt sustainable practices and improve their livelihoods.

The story of the Green Revolution is one of innovation and adaptation. It highlights the potential of human ingenuity to address pressing global challenges. While the initial phases of the revolution focused on maximizing production, the subsequent evolution underscored the importance of sustainability and inclusivity in agriculture. By learning from past experiences and embracing new technologies, the agricultural sector continues to evolve, striving to meet the needs of a growing population while preserving the planet's resources for future generations.

Impact on Global Agriculture

The transformative power of the Green Revolution on global agriculture is profound and far-reaching. During the mid-20th century, agriculture underwent a seismic shift, fundamentally altering the way crops were grown and distributed worldwide. This period marked a significant leap in agricultural productivity, spurred by the introduction of high-yielding crop varieties, advanced irrigation techniques, and the widespread use of chemical fertilizers and pesticides.

One of the most immediate and visible impacts of the Green Revolution was the dramatic increase in food production. Countries such as India, which faced the specter of famine, witnessed unprecedented growth in cereal production. The introduction of high-yield variety (HYV) seeds, particularly for staple crops like wheat and rice, played a crucial role in this agricultural boom. These seeds, engineered to maximize output, were capable of producing significantly higher yields than traditional varieties. The result was a substantial reduction in food shortages and a shift from food scarcity to self-sufficiency in many developing nations.

This surge in food production had far-reaching socio-economic consequences. In regions where food insecurity was once a pressing concern, the Green Revolution helped stabilize economies and improve living standards. Increased agricultural output allowed countries to reduce their dependence on food imports, bolstering national food security. This newfound stability enabled governments to redirect resources towards other critical areas of development, such as education, healthcare, and infrastructure.

The social fabric of rural communities also experienced change as a result of the Green Revolution. With the mechanization of farming practices and the introduction of modern agricultural techniques, the labor requirements for traditional farming decreased. This shift led to a migration of rural populations towards urban centers in search of new employment opportunities. While this urban migration contributed to economic growth in cities, it also posed challenges in terms of urban planning and infrastructure development.

The Green Revolution's impact on global agriculture was not without its challenges. The widespread adoption of chemical fertilizers and pesticides raised environmental concerns. The intensive use of these inputs often led to soil degradation, water pollution, and the depletion of natural resources. As a result, the environmental sustainability of the practices introduced during the Green Revolution came under scrutiny. The reliance on monoculture, or the cultivation of a single crop over large areas, further exacerbated these issues by reducing biodiversity and increasing vulnerability to pests and diseases.

Another significant impact was the disparity in the distribution of benefits. While wealthier farmers with access to resources and infrastructure thrived, smallholder and subsistence farmers often struggled to compete. The cost of acquiring high-yield seeds, fertilizers, and machinery was prohibitive for many, leading to a widening gap between large-scale commercial farmers and smaller, traditional ones. This inequality highlighted the need for more inclusive strategies to ensure that all farmers could benefit from advancements in agricultural technology.

The Green Revolution also had implications for global trade. As countries became more self-sufficient in food production, the dynamics of international trade shifted. Exporting nations faced new challenges as demand for imported food declined in countries that had achieved agricultural self-sufficiency. This change prompted a reevaluation of trade policies and agricultural subsidies, as nations sought to balance domestic food security with global economic interests.

In light of these challenges, the Green Revolution served as a catalyst for the development of sustainable agricultural practices. The need to address environmental concerns and social inequalities led to the exploration of alternative farming methods that prioritized ecological balance and social equity. Practices such as organic farming, agroecology, and permaculture emerged as viable alternatives, aiming to create a more sustainable and resilient global food system.

The legacy of the Green Revolution continues to influence agricultural policy and research. Governments, international organizations, and research institutions are increasingly focused on developing technologies that enhance productivity while minimizing environmental impact. Innovations such as precision agriculture, which uses technology to optimize resource use, and climate-smart agriculture, which aims to adapt farming practices to changing climate conditions, are at the forefront of this evolution.

Key Players and Innovations

The Green Revolution was a monumental chapter in the history of agriculture, driven by a consortium of visionary scientists, policymakers, and institutions who were dedicated to transforming agricultural practices to meet the demands of a burgeoning global population. These key players, through their innovative approaches and relentless efforts, catalyzed a revolution that would forever alter the landscape of food production.

Norman Borlaug, often hailed as the "Father of the Green Revolution," stands as a towering figure in this transformative era. His pioneering work in developing high-yielding and disease-resistant wheat varieties was instrumental in averting widespread famine in countries like Mexico, India, and Pakistan. Borlaug's relentless pursuit of agricultural research and his commitment to alleviating hunger earned him the Nobel Peace Prize in 1970, solidifying his legacy as a key architect of the Green Revolution.

Another significant player was M.S. Swaminathan, an Indian geneticist and international administrator renowned for his leadership in introducing and furthering the Green Revolution in India. Swaminathan's collaboration with Borlaug and his implementation of new wheat and rice varieties were critical in transforming India's agricultural landscape from a state of chronic food shortage to one of self-sufficiency. His efforts not only improved crop yields but also laid the groundwork for future agricultural research and policy in India.

The Rockefeller Foundation and the Ford Foundation played pivotal roles in funding and supporting agricultural research during this period. By establishing research centers and providing resources for the development and dissemination of high-yielding crop varieties, these institutions were instrumental in facilitating the widespread adoption of Green Revolution technologies. Their commitment to agricultural development helped bridge the gap between research and practical application, ensuring that innovations reached farmers on the ground.

The International Rice Research Institute (IRRI) in the Philippines emerged as a crucial hub for rice research and development. The institute's work in developing IR8, a high-yielding rice variety also known as "Miracle Rice," had a profound impact on rice production in Asia. IR8's success in increasing yield and resistance to pests exemplified the potential of scientific innovation to address food shortages and improve food security in rice-dependent regions.

In addition to these key players, local governments and agricultural ministries played a vital role in implementing and scaling Green Revolution technologies. By creating supportive policies, investing in infrastructure, and facilitating access to credit and resources, these institutions helped ensure the successful adoption of new agricultural practices. Their efforts were crucial in creating an enabling environment that allowed farmers to embrace change and improve their productivity.

The innovations that emerged during the Green Revolution were diverse and multifaceted. High-yielding variety (HYV) seeds were at the forefront of this transformation, engineered to produce more grain per plant and exhibit resistance to diseases and adverse environmental conditions. These seeds, combined with the increased use of chemical fertilizers and pesticides, significantly boosted crop yields and helped stabilize food supplies in many regions.

Irrigation systems were another critical innovation, addressing the limitations posed by traditional rain-fed agriculture. The development and expansion of irrigation infrastructure enabled farmers to cultivate crops in arid and semi-arid regions, reducing their reliance on unpredictable rainfall. This advancement not only increased agricultural productivity but also allowed for multiple cropping cycles, further enhancing food production.

The mechanization of farming practices, through the introduction of tractors, threshers, and other machinery, revolutionized traditional agricultural methods. By reducing the reliance on manual labor, mechanization increased efficiency and productivity, allowing farmers to cultivate larger areas of land with fewer resources. This shift not only improved crop yields but also contributed to the modernization of rural communities.

The Green Revolution also saw the emergence of integrated pest management (IPM) strategies, which aimed to reduce the reliance on chemical pesticides by incorporating biological, cultural, and mechanical control methods. By promoting a holistic approach to pest management, these strategies helped mitigate the environmental impact of pesticide use and contributed to the sustainability of farming practices.

While the innovations of the Green Revolution were groundbreaking, they were not without challenges. The heavy reliance on chemical inputs raised environmental concerns, as overuse led to soil degradation, water pollution, and a reduction in biodiversity. The monoculture practices promoted during this era often resulted in the loss of traditional crop varieties and a decrease in genetic diversity. These issues underscored the need for a more balanced approach to agricultural development, one that prioritized sustainability and ecological health.

In response to these challenges, a new wave of innovation began to emerge, focusing on sustainable farming practices that integrated traditional knowledge with modern scientific advancements. Agroecology, organic farming, and permaculture gained traction as viable alternatives, offering environmentally friendly methods for enhancing soil fertility, conserving water, and promoting biodiversity.

The role of technology in modern agriculture continues to evolve, with precision agriculture and climate-smart practices at the forefront of contemporary innovation. By leveraging data and technology, farmers can optimize resource use, monitor crop health, and adapt to changing climate conditions, ensuring the resilience and sustainability of agricultural systems for future generations.

Challenges and Criticisms

The Green Revolution, while heralded as a monumental achievement in increasing agricultural productivity, faced a myriad of challenges and criticisms that have sparked ongoing debates among scholars, policymakers, and environmentalists. As with any major transformation, the benefits came with consequences, and the intricacies of these outcomes provide a rich tapestry of lessons for the future of agriculture.

One of the foremost criticisms of the Green Revolution was its environmental impact. The rapid increase in agricultural output was largely driven by the extensive use of chemical fertilizers and pesticides. Although these inputs played a vital role in boosting crop yields, their overuse led to significant environmental degradation. Soils became depleted of essential nutrients, and their structure was compromised. The widespread application of pesticides resulted in the contamination of water bodies, affecting aquatic life and ecosystems. Furthermore, the reliance on chemical inputs contributed to a reduction in biodiversity, as traditional crop varieties were often replaced by high-yield monocultures.

Monoculture practices, a hallmark of the Green Revolution, posed another significant challenge. By focusing on a limited number of high-yield crops such as wheat and rice, genetic diversity within agricultural systems was reduced. This lack of diversity made crops more vulnerable to pests, diseases, and changing environmental conditions. The erosion of traditional agricultural knowledge and practices further exacerbated this issue, as indigenous crop varieties, which were often more resilient to local conditions, were replaced by standardized high-yield varieties.

The social implications of the Green Revolution were equally profound. While it succeeded in increasing food production and reducing hunger in many regions, the distribution of its benefits was uneven. Large-scale commercial farmers with access to resources and infrastructure were able to capitalize on the new technologies, while smallholder and subsistence farmers often struggled to keep pace. The high costs associated with acquiring seeds, fertilizers, and machinery created barriers for poorer farmers, leading to increased socio-economic disparities within rural communities.

The mechanization of agriculture, another key component of the Green Revolution, resulted in the displacement of rural laborers. As machines began to replace manual labor, many farmworkers found themselves without employment, prompting mass migrations to urban centers in search of work. This urban influx contributed to overcrowding and strained infrastructure in cities, creating new socio-economic challenges.

The Green Revolution's focus on staple crops, while instrumental in addressing caloric deficiencies, often came at the expense of nutritional diversity. The increased emphasis on wheat and rice overshadowed the cultivation of nutrient-rich traditional crops, impacting dietary diversity and contributing to hidden hunger or micronutrient deficiencies. This nutritional imbalance highlighted the need for a more holistic approach to food production that considers both quantity and quality.

The economic ramifications of the Green Revolution were complex and multifaceted. While some countries benefited from increased food self-sufficiency and reduced dependence on imports, others faced challenges in adapting to changing trade dynamics. Export-oriented economies had to navigate fluctuations in global demand as countries that had achieved self-sufficiency reduced their reliance on imports. This shift required a reevaluation of trade policies and agricultural subsidies, as nations sought to balance domestic food security with international economic interests.

Critics of the Green Revolution also pointed to the lack of consideration for local cultural and social contexts in the implementation of new agricultural practices. The top-down approach often failed to account for the unique needs and circumstances of different regions and communities, leading to resistance and skepticism among some farmers. This disconnect underscored the importance of involving local stakeholders in the decision-making process and adapting technologies to fit local conditions and preferences.

In response to these challenges and criticisms, there has been a growing movement towards more sustainable and inclusive agricultural practices. The emphasis has shifted from maximizing production at all costs to balancing productivity with environmental stewardship and social equity. This evolution has led to the exploration of alternative farming methods that prioritize ecological health and community well-being.

Agroecology, organic farming, and permaculture have gained traction as viable alternatives to traditional Green Revolution practices. These approaches seek to create resilient agricultural systems by promoting biodiversity, enhancing soil health, and reducing reliance on chemical inputs. By integrating traditional knowledge with modern scientific advancements, these practices offer a path towards more sustainable food production.

Water conservation has also become a focal point in addressing the environmental challenges of the Green Revolution. The development of efficient irrigation techniques, such as drip irrigation and rainwater harvesting, has helped optimize water use and reduce wastage. These innovations not only conserve valuable resources but also improve crop yields by ensuring that plants receive adequate moisture.

The lessons learned from the Green Revolution have also influenced agricultural policy and research. There is a growing recognition of the need to support smallholder farmers and ensure that they have access to the resources and knowledge necessary to adopt sustainable practices. Initiatives aimed at providing credit, training, and infrastructure have been launched to empower these farmers and improve their livelihoods.

Future Prospects and Trends

Envisioning the future of agriculture involves navigating a complex landscape where innovation, sustainability, and resilience converge. The trajectory of farming practices has always been shaped by the pressing needs of society, and as we move forward, these needs are becoming increasingly intertwined with global challenges such as climate change, population growth, and resource scarcity. The future prospects and trends in agriculture are thus framed by the necessity to produce more with less, all while preserving the ecological balance of our planet.

One of the most promising trends reshaping agriculture is the rise of precision farming. This approach leverages technology to optimize crop production by using data-driven insights to inform decisions at every stage of the farming process. With tools like GPS mapping, drones, and remote sensing, farmers can monitor field conditions in real-time, allowing them to apply water, fertilizers, and pesticides with pinpoint accuracy. This not only enhances productivity but also reduces waste and environmental impact. Through precision farming, the agricultural sector is poised to become more efficient, resilient, and sustainable.

The integration of digital technologies is further transforming the agricultural landscape. From blockchain for supply chain transparency to the Internet of Things (IoT) for real-time monitoring, digital solutions are enhancing connectivity and data sharing across the agricultural value chain. These technologies facilitate better resource management, improve traceability, and empower farmers with the information needed to make informed decisions. As digital infrastructure continues to develop, the potential for data-driven agriculture to revolutionize food production becomes increasingly attainable.

Climate-smart agriculture is another critical trend that addresses the dual challenges of increasing productivity and mitigating climate change impacts. This approach involves adopting practices that enhance the resilience of agricultural systems to climate variability while reducing greenhouse gas emissions. Techniques such as agroforestry, conservation tillage, and crop diversification are being explored to create more robust farming systems that can withstand extreme weather events and shifting climatic conditions. By integrating climate considerations into agricultural planning, farmers can contribute to global efforts to combat climate change while safeguarding their livelihoods.

The future of agriculture also necessitates a shift towards regenerative practices that focus on restoring and enhancing the health of ecosystems. Regenerative agriculture emphasizes the importance of soil health, biodiversity, and ecological balance. Practices such as cover cropping, rotational grazing, and composting not only improve soil fertility but also increase carbon sequestration and support biodiversity. Regenerative approaches offer a pathway to building resilient agricultural systems that can adapt to environmental changes while contributing to the restoration of natural ecosystems.

Urban agriculture is emerging as a viable solution to address food security challenges in rapidly growing urban areas. By bringing food production closer to consumers, urban farming reduces the need for transportation and minimizes food waste. Techniques such as vertical farming, hydroponics, and aquaponics are being utilized to grow fresh produce in urban environments, utilizing space efficiently and sustainably. Urban agriculture not only provides fresh and nutritious food but also offers opportunities for community engagement and education, fostering a closer connection between people and their food sources.

The role of biotechnology in shaping the future of agriculture cannot be overlooked. Advances in genetic engineering and synthetic biology hold the potential to develop crops that are more resilient to pests, diseases, and environmental stresses. These innovations can enhance nutritional content, reduce the need for chemical inputs, and improve crop yields. However, the adoption of biotechnology in agriculture also raises ethical and regulatory considerations that must be carefully navigated to ensure that the benefits are realized equitably and sustainably.

Consumer preferences are increasingly influencing agricultural trends, as there is a growing demand for transparency, sustainability, and ethical production. Consumers are becoming more conscious of the environmental and social impacts of their food choices, driving a shift towards organic, locally-sourced, and sustainably-produced foods. This trend is prompting producers to adopt practices that align with consumer values, creating opportunities for niche markets and value-added products.

Policy and governance will play a crucial role in shaping the future of agriculture. As governments and international organizations recognize the importance of sustainable food systems, there is a growing emphasis on creating supportive policies and frameworks that promote innovation and resilience. Investment in research and development, infrastructure, and education will be vital to ensure that farmers, particularly smallholders, have access to the resources and knowledge needed to thrive in a changing world.

Collaboration and partnerships will also be key in addressing the complex challenges facing agriculture. By fostering collaboration between governments, research institutions, private sector actors, and civil society, it is possible to create synergies that drive innovation and support sustainable development. Multi-stakeholder approaches can facilitate the sharing of knowledge, resources, and expertise, enabling the agricultural sector to adapt and evolve in response to emerging trends and challenges.

Chapter 2

The Foundations of Sustainable Farming

Principles of Sustainability

Sustainability in agriculture is a guiding principle that seeks to harmonize the need for food production with the imperative to preserve the environment and enhance social equity. At its core, sustainability in farming involves practices that are ecologically sound, economically viable, and socially responsible, ensuring that the needs of present and future generations can be met without compromising the health of our planet.

One of the fundamental principles of sustainability in agriculture is the conservation of natural resources. This involves the careful management of soil, water, and biodiversity to maintain ecological balance and prevent degradation. Soil health is a critical component, as it provides the foundation for plant growth and nutrient cycling. Practices such as crop rotation, cover cropping, and reduced tillage help maintain soil fertility and structure, reduce erosion, and increase organic matter content. By preserving soil health, farmers can ensure long-term productivity and resilience against environmental stresses.

Water management is equally vital in sustainable agriculture. Efficient irrigation techniques, such as drip irrigation and rainwater harvesting, optimize water use and minimize wastage. These practices not only conserve water resources but also enhance crop yields by delivering moisture directly to plant roots. Protecting water quality is another aspect of sustainability, achieved by minimizing the use of chemical fertilizers and pesticides that can leach into waterways. Buffer strips and constructed wetlands are examples of strategies used to filter runoff and protect aquatic ecosystems.

Biodiversity conservation is another cornerstone of sustainable farming. Diverse agricultural systems are more resilient to pests, diseases, and climate variability. Integrating a variety of crops, livestock, and natural habitats within agricultural landscapes can support beneficial organisms, enhance pollination, and provide habitat for wildlife. Agroforestry, which combines trees with crops or livestock, is an example of a practice that promotes biodiversity while offering additional benefits such as carbon sequestration and soil stabilization.

Economic viability is a critical aspect of sustainability, as farmers must be able to maintain their livelihoods while adopting sustainable practices.

This requires access to markets, fair prices, and financial resources to invest in sustainable technologies and practices. Diversification of income sources, such as through agro-tourism or value-added products, can enhance economic resilience. Supporting local food systems and shortening supply chains can also contribute to economic sustainability by reducing transportation costs and increasing market access for small-scale farmers.

Social responsibility is an integral part of sustainable agriculture, emphasizing the well-being of farming communities and equitable access to resources. This includes fair labor practices, safe working conditions, and opportunities for education and training. Empowering women and marginalized groups in agriculture is essential for achieving social equity and fostering inclusive development. Community engagement and participatory approaches to decision-making ensure that the diverse needs and perspectives of stakeholders are considered in agricultural planning.

Sustainable agriculture also involves the integration of traditional knowledge and modern science. Indigenous practices, honed over generations, offer valuable insights into effective resource management and ecological balance. By combining these practices with scientific research and innovation, farmers can develop adaptive strategies that enhance sustainability. This knowledge exchange fosters a holistic understanding of agricultural systems and their interactions with the environment.

Climate change is a significant challenge that underscores the need for sustainable agricultural practices. Farmers must adapt to changing weather patterns, increased frequency of extreme events, and shifting growing seasons. Climate-smart agriculture, which incorporates adaptive measures such as drought-resistant crop varieties and improved water management, is a response to these challenges. By reducing greenhouse gas emissions and increasing carbon sequestration, sustainable practices also contribute to climate change mitigation.

Consumer awareness and demand for sustainable products are driving changes in agricultural practices. As people become more conscious of the environmental and social impacts of their food choices, they seek products that align with their values. This shift is prompting producers to adopt practices that prioritize sustainability, transparency, and ethical production. Certification schemes and eco-labels are tools that help consumers identify sustainable products and encourage producers to meet rigorous standards.

Policy and governance play a crucial role in promoting sustainability in agriculture. Supportive policies and incentives can encourage the adoption of sustainable practices and provide the necessary resources for farmers to transition. Investment in research and development, infrastructure, and education is essential to advance sustainable agriculture and build capacity within farming communities. Collaborative efforts between governments, research institutions, and civil society can create a conducive environment for sustainable development.

Education and knowledge sharing are vital components of sustainability, empowering farmers with the skills and information needed to implement sustainable practices. Extension services, farmer field schools, and peer-to-peer networks facilitate the dissemination of knowledge and encourage innovation. By fostering a culture of learning and collaboration, agricultural communities can adapt to emerging challenges and opportunities.

The Role of Biodiversity

Biodiversity, often referred to as the variety of life on Earth, plays a crucial role in the stability and sustainability of ecosystems, including agricultural systems. The intricate interplay between different species and their environments creates a web of interactions that supports essential ecosystem functions such as nutrient cycling, pollination, pest control, and climate regulation. In the context of agriculture, biodiversity is not merely a background feature but a vital component that enhances productivity, resilience, and sustainability.

In traditional agricultural systems, biodiversity was naturally integrated through practices like polycultures and crop rotations. Farmers cultivated multiple species and varieties of crops, which not only provided diverse sources of food and income but also helped maintain soil fertility and reduce pest and disease pressures. This diversity acted as a buffer against environmental fluctuations, ensuring that if one crop failed due to adverse conditions, others might still thrive. The genetic diversity within crop varieties also contributed to resilience, as certain varieties might possess resistance to specific pests or diseases.

Pollinators, such as bees, butterflies, and birds, are a prime example of biodiversity's indispensable role in agriculture. These creatures facilitate the reproduction of many flowering plants, including a significant portion of the world's food crops. Without them, the yields of fruits, nuts, and vegetables would plummet, leading to decreased food availability and increased prices. Maintaining habitats for pollinators, such as wildflower strips and hedgerows, is thus critical for sustaining agricultural productivity.

Biodiversity also contributes to natural pest control. Predators and parasitoids, which are part of the wider ecological community, help keep pest populations in check. When agricultural systems lack diversity, they often become more vulnerable to pest outbreaks, leading to increased reliance on chemical pesticides. By fostering a diverse ecosystem that includes beneficial insects, birds, and other predators, farmers can reduce their dependence on synthetic inputs and promote more sustainable pest management strategies.

Soil biodiversity, comprising a myriad of microorganisms, fungi, and invertebrates, is fundamental to soil health and fertility. These organisms decompose organic matter, recycle nutrients, and improve soil structure, enhancing its ability to retain water and support plant growth. Practices that maintain or enhance soil biodiversity, such as reduced tillage, organic amendments, and cover cropping, are essential for building resilient agricultural systems capable of withstanding climatic extremes.

The erosion of agricultural biodiversity, driven by the widespread adoption of monocultures and the focus on high-yielding varieties, poses significant risks to food security and ecosystem health. This loss of diversity reduces the genetic pool available for crop improvement and adaptation, making food systems more vulnerable to pests, diseases, and changing environmental conditions. To counteract this trend, there is a growing movement towards conserving and utilizing traditional crop varieties and wild relatives, which hold valuable genetic traits for resistance and resilience.

Agroecology, which emphasizes the integration of biodiversity into farming systems, offers a framework for harnessing the benefits of biodiversity in agriculture. By designing diversified systems that mimic natural ecosystems, agroecology seeks to optimize the interactions between plants, animals, and microorganisms to enhance productivity and sustainability. Techniques such as intercropping, agroforestry, and permaculture exemplify how biodiversity can be leveraged to create resilient agricultural landscapes.

The role of biodiversity extends beyond farm boundaries to encompass the broader landscape and its ecological processes. Connectivity between habitats, through ecological corridors and buffer zones, supports the movement of species and the flow of ecosystem services across the landscape. Landscape-level planning that incorporates biodiversity conservation can enhance ecosystem resilience and contribute to broader conservation goals.

Biodiversity also holds cultural and economic value for communities. Many traditional farming systems are rooted in cultural practices that have evolved over generations, reflecting a deep understanding of local biodiversity and its uses. Protecting and revitalizing this traditional knowledge is crucial for sustaining cultural heritage and fostering innovation in sustainable agriculture.

Engagement with local communities and stakeholders is essential for the successful integration of biodiversity into agricultural systems. Participatory approaches that involve farmers, indigenous peoples, and local communities in decision-making processes can ensure that biodiversity conservation efforts are contextually relevant and culturally appropriate. Empowering communities to manage and conserve their biodiversity resources can lead to more effective and sustainable outcomes.

Policy and governance frameworks play a critical role in supporting biodiversity conservation in agriculture. Incentives for sustainable practices, investment in research and development, and the establishment of protected areas can create an enabling environment for biodiversity-friendly agriculture. International agreements and collaborations, such as the Convention on Biological Diversity (CBD), provide platforms for countries to work together towards common conservation goals.

Education and awareness-raising are vital for fostering a deeper understanding of the importance of biodiversity in agriculture. By promoting biodiversity literacy among farmers, consumers, and policymakers, we can create a more informed society that values and supports the conservation of biodiversity. Educational programs, outreach initiatives, and media campaigns can help bridge knowledge gaps and inspire action at all levels.

Research and innovation are key to advancing our understanding of biodiversity's role in agriculture and developing new strategies for its conservation and use. Interdisciplinary research that integrates ecology, agronomy, and social sciences can provide insights into the complex interactions between biodiversity and agricultural systems. Collaborative research efforts that involve farmers and local communities can ensure that scientific findings are translated into practical applications.

Soil Health and Management

Soil health is the cornerstone of sustainable agriculture, providing the foundation upon which plants grow and ecosystems thrive. Understanding and managing soil health requires a holistic approach, focusing on the biological, chemical, and physical properties that contribute to its overall functionality. Healthy soil supports plant growth, stores carbon, filters water, and sustains diverse microorganisms, making it an indispensable resource for food production and environmental health.

The biological aspect of soil health is characterized by the myriad of organisms that inhabit the soil, including bacteria, fungi, earthworms, and insects. These organisms form a complex food web, breaking down organic matter and recycling nutrients essential for plant growth. Beneficial microorganisms, such as mycorrhizal fungi, form symbiotic relationships with plant roots, enhancing nutrient uptake and increasing plant resilience to environmental stresses. To support soil biology, farmers can implement practices such as adding organic matter, reducing chemical inputs, and promoting biodiversity through crop rotations and cover cropping.

Chemically, soil health is determined by the availability of essential nutrients and the balance of soil pH. Nutrient-rich soils provide plants with the necessary elements for growth, including nitrogen, phosphorus, and potassium. However, excessive use of chemical fertilizers can lead to nutrient imbalances, soil acidification, and environmental pollution. Sustainable nutrient management involves regular soil testing to determine nutrient needs, applying fertilizers based on soil test results, and incorporating organic amendments such as compost and green manure to improve nutrient availability and soil structure.

The physical properties of soil, including texture, structure, and porosity, influence water retention, root penetration, and air exchange. Soils with good structure have stable aggregates that allow for adequate water infiltration and drainage, reducing the risk of erosion and waterlogging. Compaction, often caused by heavy machinery or overgrazing, can damage soil structure, leading to reduced porosity and impaired root growth. Practices such as reduced tillage, controlled traffic farming, and maintaining ground cover can help preserve soil structure and prevent compaction.

Organic matter, a key component of soil health, improves soil structure, enhances nutrient retention, and increases water-holding capacity. It serves as a food source for soil organisms and contributes to the formation of stable soil aggregates. Increasing soil organic matter can be achieved through the addition of crop residues, animal manures, and compost, as well as by implementing practices like cover cropping and agroforestry. These strategies not only enhance soil health but also contribute to climate change mitigation by sequestering carbon in the soil.

Soil erosion is a significant threat to soil health, leading to the loss of fertile topsoil and associated nutrients. Erosion can result from wind, water, and tillage, and its impacts are exacerbated by deforestation, overgrazing, and poor land management. To combat erosion, farmers can adopt conservation tillage practices that minimize soil disturbance, establish vegetative buffers to slow water flow, and use contour farming and terracing to reduce runoff. By protecting the soil surface and maintaining cover, these practices help retain soil and its valuable nutrients.

Water management is intimately linked to soil health, as the ability of soil to retain and supply water affects plant growth and resilience to drought. Well-structured soils with high organic matter content have greater water-holding capacity, reducing the need for irrigation and improving water-use efficiency. Practices such as mulching, drip irrigation, and rainwater harvesting can further enhance water conservation and distribution, ensuring that crops receive adequate moisture while minimizing water wastage.

Crop diversity is another important factor in soil health management. Diverse cropping systems, including intercropping and crop rotations, disrupt pest and disease cycles, enhance nutrient cycling, and promote beneficial soil organisms. Rotating crops with different rooting depths and nutrient requirements can improve soil structure and fertility, while cover crops provide continuous soil cover and contribute to organic matter accumulation. By diversifying cropping systems, farmers can create more resilient agricultural landscapes that support healthy soils.

Monitoring and assessing soil health is essential for informed management decisions. Soil testing provides valuable information on nutrient levels, pH, organic matter content, and biological activity, helping farmers tailor their management practices to meet the specific needs of their soils. Visual assessments, such as examining soil structure, color, and presence of earthworms, can also provide insights into soil health. Regular monitoring allows for the early detection of potential issues and the implementation of corrective measures to maintain or improve soil health.

Education and knowledge sharing play a critical role in promoting soil health management. By learning from one another and sharing best practices, farmers can develop innovative solutions to soil health challenges. Extension services, farmer field schools, and peer networks provide valuable platforms for knowledge exchange and capacity building, empowering farmers to adopt and adapt soil health practices that suit their unique contexts.

Policy and governance frameworks are essential for supporting soil health initiatives. Creating incentives for sustainable soil management, investing in research and development, and providing access to resources and training can facilitate the widespread adoption of soil health practices. Collaborative efforts between governments, research institutions, and farmers are necessary to develop and implement policies that promote long-term soil stewardship.

Water Conservation Techniques

Water, the lifeblood of agriculture, is an increasingly precious resource, and its conservation is paramount as we strive for sustainable farming practices. Ensuring that water is used efficiently not only supports crop growth and food production but also preserves ecosystems and mitigates the impacts of climate change. Implementing water conservation techniques in agriculture requires a combination of traditional wisdom and modern innovation, tailored to the specific needs and conditions of each farming landscape.

The first step in conserving water is understanding the water requirements of different crops and aligning irrigation practices accordingly. Different plants have varying water needs, which can change throughout their growth stages. By closely monitoring these needs, farmers can optimize water use and reduce waste. This approach involves assessing local climate conditions, soil types, and crop characteristics to develop a tailored irrigation schedule that maximizes efficiency.

Drip irrigation is one of the most effective water-saving technologies available to farmers. This method delivers water directly to the root zone of plants through a network of tubes and emitters, minimizing evaporation and runoff. By providing precise amounts of water exactly where it's needed, drip irrigation can significantly reduce water use while promoting healthy crop growth. Additionally, it allows for the simultaneous application of fertilizers, enhancing nutrient uptake and further improving efficiency.

Rainwater harvesting is another valuable technique for water conservation, capturing and storing rainfall for use during dry periods. This approach can be as simple as collecting runoff from rooftops or as complex as constructing reservoirs and ponds to store large quantities of water. By utilizing rainwater, farmers can supplement their irrigation needs without drawing from local water sources, reducing their environmental impact and increasing resilience to drought.

Mulching, the practice of covering soil with organic or inorganic materials, plays a crucial role in conserving soil moisture. Mulch acts as a protective barrier, reducing evaporation, suppressing weed growth, and moderating soil temperature. Organic mulches, such as straw, wood chips, and leaves, also improve soil structure and fertility as they decompose. By retaining moisture in the soil, mulching can reduce the need for frequent irrigation and support plant health during dry spells.

The adoption of conservation tillage practices, such as no-till or reduced tillage, can also enhance water conservation. These methods minimize soil disturbance, preserving soil structure and organic matter content. Undisturbed soils have better water infiltration and retention capabilities, which can reduce surface runoff and increase the availability of water to plants. By maintaining a protective cover of crop residues on the soil surface, conservation tillage helps conserve moisture and prevent erosion.

Crop diversification and rotation are additional strategies that can contribute to water conservation. By growing a variety of crops with different water needs and rooting depths, farmers can optimize water use and reduce pressure on specific water resources. Deep-rooted plants, for example, can access water stored deeper in the soil profile, reducing the need for supplementary irrigation. Rotating crops also helps maintain soil health and structure, further supporting efficient water use.

Implementing agroforestry systems, which integrate trees with crops or livestock, can enhance water conservation through multiple mechanisms. Trees improve water infiltration and reduce runoff, while their roots stabilize soil and prevent erosion. The shade provided by tree canopies can also lower soil temperatures and reduce evaporation, conserving moisture for underlying crops. Agroforestry systems create a more diverse and resilient agricultural landscape, capable of withstanding water scarcity.

Efficient water management is not limited to the farm scale; landscape-level approaches are essential for comprehensive water conservation. Watershed management involves coordinating water use and conservation efforts across entire catchment areas, recognizing the interconnectedness of water resources. By fostering collaboration between stakeholders, including farmers, communities, and policymakers, watershed management can optimize water allocation and enhance the sustainability of water use.

Technological advancements are also playing a pivotal role in water conservation. Remote sensing, soil moisture sensors, and weather forecasting tools provide farmers with real-time data on soil and atmospheric conditions, enabling them to make informed irrigation decisions. These technologies facilitate precision agriculture, where water and other inputs are applied with accuracy and efficiency, minimizing waste and maximizing productivity.

Education and capacity building are integral to the successful adoption of water conservation techniques. By raising awareness and providing training on sustainable water management practices, farmers can develop the skills and knowledge needed to implement these techniques effectively. Extension services, farmer field schools, and peer-to-peer learning networks offer valuable platforms for sharing experiences and best practices, fostering a culture of innovation and sustainability.

Policy and governance frameworks play a critical role in supporting water conservation initiatives. Governments and institutions can create enabling environments by providing incentives for sustainable practices, investing in research and development, and ensuring equitable access to water resources. Collaborative efforts at local, national, and international levels are necessary to address the complex challenges of water scarcity and ensure the long-term sustainability of agricultural systems.

Consumer awareness and demand for sustainably produced food can also drive water conservation efforts. As consumers become more conscious of the environmental impacts of their food choices, they are increasingly seeking products that prioritize sustainability. By supporting producers who implement water-saving practices, consumers can contribute to the broader goals of water conservation and sustainable development.

Energy Efficiency in Agriculture

Energy efficiency in agriculture is a critical component of sustainable farming practices, offering a pathway to reduce costs, minimize environmental impacts, and enhance the resilience of agricultural systems. As the demand for food continues to rise alongside concerns about climate change and resource depletion, optimizing energy use on farms is more important than ever. By examining the energy inputs and outputs of agricultural operations, farmers can identify opportunities to conserve energy, reduce greenhouse gas emissions, and improve overall productivity.

At the heart of energy efficiency in agriculture is the concept of energy audits. These assessments help farmers understand their energy consumption patterns, identify inefficiencies, and develop strategies to improve energy use. An energy audit typically involves evaluating the farm's operations, equipment, and infrastructure to pinpoint areas where energy is being wasted or where improvements can be made. By implementing the recommendations from an energy audit, farmers can often achieve significant energy savings and cost reductions.

Renewable energy sources, such as solar, wind, and biomass, offer promising alternatives to traditional fossil fuels in agriculture. Solar panels can be installed on rooftops or open fields to generate electricity for farm operations, reducing reliance on grid power and lowering energy costs. Wind turbines can harness the energy of the wind to produce electricity, especially in areas with consistent wind patterns. Biomass energy, derived from organic materials such as crop residues and animal waste, can be converted into biogas or biofuels to power farm machinery and equipment. By integrating renewable energy sources, farmers not only reduce their carbon footprint but also increase their energy independence and resilience.

Improving the efficiency of farm machinery and equipment is another key aspect of energy conservation. Regular maintenance, such as cleaning and lubricating equipment, can enhance performance and extend the lifespan of machinery. Upgrading to more energy-efficient models, such as tractors with advanced fuel efficiency technologies, can further reduce energy consumption and emissions. Precision agriculture technologies, which use GPS and sensors to optimize field operations, can also play a significant role in improving energy efficiency. By applying inputs such as fertilizers and pesticides with precision, farmers can reduce the energy required for application and minimize waste.

In irrigation, energy efficiency can be achieved through the adoption of modern, efficient systems and practices. Drip irrigation, as mentioned earlier, delivers water directly to plant roots with minimal energy input compared to traditional flood irrigation methods. Variable speed pumps, which adjust their output based on the water demand, can significantly reduce energy use in irrigation systems. Monitoring soil moisture levels and weather conditions can help farmers irrigate only when necessary, conserving both water and energy.

The energy efficiency of buildings and infrastructure on farms also presents opportunities for conservation. Proper insulation and ventilation in barns, greenhouses, and storage facilities can reduce the need for heating and cooling, resulting in lower energy consumption. Energy-efficient lighting, such as LED bulbs, can provide adequate illumination with less energy than traditional lighting options. Implementing energy-efficient designs and materials in new construction projects can further enhance the sustainability of farm operations.

Transportation is another area where energy efficiency can be improved. Optimizing the logistics of transporting goods, such as by reducing travel distances and maximizing load capacities, can lower fuel consumption and emissions. Transitioning to vehicles powered by alternative fuels, such as electric or biodiesel, can also contribute to energy savings and a reduced environmental impact.

Education and training are essential for promoting energy efficiency in agriculture. Farmers and farmworkers need access to information and resources that enable them to implement energy-saving practices effectively. Workshops, extension services, and online resources can provide valuable guidance on energy management strategies and technologies. By fostering a culture of innovation and sustainability, the agricultural community can continue to develop new approaches to energy efficiency.

Policy and governance play a crucial role in supporting energy efficiency initiatives in agriculture. Governments can offer incentives, such as tax credits and grants, to encourage the adoption of energy-efficient technologies and practices. Research and development funding can support the creation and dissemination of innovative solutions for energy conservation. Collaborative efforts between policymakers, research institutions, and the agricultural sector are necessary to create an enabling environment for energy efficiency.

Consumer awareness and demand for sustainably produced food products can also drive energy efficiency efforts in agriculture. As consumers become more conscious of the environmental impacts of their food choices, they increasingly seek products that prioritize sustainability. By supporting producers who implement energy-efficient practices, consumers can contribute to the broader goals of energy conservation and sustainable development.

Chapter 3

Crop Management Innovations

Precision Agriculture: Data-Driven Decisions

Precision agriculture represents a transformative approach in modern farming, leveraging data and technology to enhance decision-making processes and optimize agricultural outputs. By integrating advanced tools such as GPS, remote sensing, and data analytics, precision agriculture allows farmers to manage their fields with unprecedented accuracy, tailoring practices to the specific needs of each crop and plot of land. This data-driven approach not only boosts productivity but also contributes to sustainability by minimizing resource use and environmental impact.

At the core of precision agriculture is the collection and analysis of data. Sensors mounted on satellites, drones, and farm machinery capture detailed information about soil conditions, crop health, weather patterns, and more. This data is then processed and analyzed to generate actionable insights, informing decisions on planting, fertilization, irrigation, and pest control. By understanding the variability within their fields, farmers can apply inputs more efficiently, targeting specific areas that require attention and avoiding overuse in others.

One of the key benefits of precision agriculture is its ability to enhance nutrient management. Soil testing and mapping technologies provide detailed information about nutrient levels across a field, allowing farmers to apply fertilizers with precision. Variable rate technology (VRT) enables the application of different rates of fertilizers, seeds, and pesticides in specific zones, maximizing their effectiveness while reducing waste and environmental runoff. This targeted approach not only improves crop yields but also helps protect water quality and reduce greenhouse gas emissions.

Irrigation management is another area where precision agriculture excels. Soil moisture sensors and weather data can be used to develop precise irrigation schedules, ensuring that crops receive the right amount of water at the right time. This reduces water waste and energy use, particularly when integrated with advanced irrigation systems like drip or pivot irrigation. By maintaining optimal soil moisture levels, farmers can improve crop health and resilience to drought, contributing to more sustainable water management practices.

Precision agriculture also plays a crucial role in pest and disease management. Remote sensing technologies, such as multispectral and hyperspectral imaging, can detect early signs of pest infestations or disease outbreaks. By identifying affected areas promptly, farmers can intervene with targeted treatments, minimizing crop loss and reducing the need for broad-spectrum pesticides. This proactive approach supports integrated pest management strategies, promoting biodiversity and reducing chemical inputs.

The integration of machine learning and artificial intelligence into precision agriculture further enhances decision-making capabilities. These technologies can process vast amounts of data quickly, identifying patterns and trends that may not be immediately apparent to human observers. Predictive models can forecast crop yields, pest outbreaks, and weather impacts, enabling farmers to make informed decisions and plan for potential challenges. By harnessing the power of AI, precision agriculture empowers farmers to optimize their operations and mitigate risks.

While the benefits of precision agriculture are considerable, successful implementation requires overcoming several challenges. Access to technology and data infrastructure can be a barrier for smallholder and resource-limited farmers. Ensuring equitable access to precision agriculture tools is essential for maximizing its potential and promoting sustainable development. Training and capacity building are also critical, as farmers need the skills and knowledge to interpret data and integrate precision technologies into their practices effectively.

Data privacy and security are important considerations in precision agriculture. With the increasing amount of data generated and shared, protecting sensitive information is paramount. Establishing clear policies and regulations on data ownership, sharing, and use can help build trust and ensure that farmers retain control over their data.

Collaboration and partnerships are key to advancing precision agriculture. By working together, farmers, technology providers, researchers, and policymakers can develop and implement solutions that meet the diverse needs of the agricultural community. Public-private partnerships and research collaborations can drive innovation and ensure that precision agriculture technologies are accessible and affordable for all farmers.

Policy frameworks play a vital role in supporting the adoption of precision agriculture. Governments can create incentives for the use of precision technologies, invest in research and development, and provide infrastructure and training support. By fostering an enabling environment, policymakers can help farmers overcome barriers and fully realize the benefits of precision agriculture.

Consumer awareness and demand for sustainably produced food can also drive the adoption of precision agriculture practices. As consumers become more conscious of the environmental and social impacts of their food choices, they increasingly seek products that prioritize sustainability and transparency. By adopting precision agriculture, farmers can meet consumer demands for responsible production, enhancing their market competitiveness and contributing to broader sustainability goals.

Integrated Pest Management

Integrated Pest Management (IPM) offers a comprehensive approach to controlling pests in agriculture by combining biological, cultural, physical, and chemical tools in a coordinated way. The goal is to manage pest populations below levels that cause economic harm while minimizing risks to human health and the environment. By understanding the life cycles and behaviors of pests, farmers can implement strategies that are both effective and sustainable.

One of the cornerstones of IPM is prevention, which involves creating conditions that are unfavorable for pest development. Crop rotation is an effective preventive measure, disrupting the life cycles of pests that specialize in certain plants. By alternating crops with different pest profiles, farmers can reduce pest populations naturally. Similarly, selecting pest-resistant crop varieties can provide a straightforward way to minimize the impact of pests without relying heavily on chemical interventions.

Cultural controls, such as proper sanitation and crop management, play a vital role in preventing pest outbreaks. Removing plant debris and weeds that harbor pests can significantly reduce the initial pest population. Additionally, adjusting planting dates and optimizing plant spacing can help crops avoid peak pest activity periods and reduce pest pressure. These simple yet effective practices can fortify crops against potential infestations.

Biological control involves using living organisms to suppress pest populations. Beneficial insects, such as ladybugs and parasitic wasps, can be introduced to target specific pests. These natural enemies help maintain the ecological balance by preying on harmful insects. Encouraging a diverse ecosystem on the farm, through practices like maintaining hedgerows and planting flowering plants, can attract and sustain these beneficial organisms, providing a natural line of defense against pests.

Physical and mechanical controls are non-chemical methods used to reduce pest populations. These include barriers such as row covers and nets, which physically block pests from reaching plants. Traps and hand-picking can also be effective, particularly in small-scale farming operations. These methods offer a direct way to manage pests without introducing chemicals into the environment.

Chemical controls are used in IPM as a last resort when other methods are insufficient to keep pest populations below damaging levels. The judicious use of pesticides involves selecting products that target specific pests while minimizing harm to non-target organisms and the environment. Farmers are encouraged to rotate pesticides with different modes of action to prevent the development of resistant pest populations. By integrating pesticides with other control measures, the overall reliance on chemicals can be reduced, leading to more sustainable pest management.

Monitoring and assessing pest populations is critical in IPM. Regular field scouting allows farmers to identify pest presence and assess damage levels. By keeping detailed records, farmers can detect trends and predict pest outbreaks, facilitating timely interventions. Threshold levels, which define the point at which pest populations will cause unacceptable damage, guide decision-making and help avoid unnecessary treatments.

Education and training are essential components of IPM. Farmers and farmworkers need to be knowledgeable about pest identification, behavior, and management strategies. Extension services, workshops, and online resources can provide valuable information and support for implementing IPM practices. By fostering a culture of continuous learning, the agricultural community can stay informed about new developments and innovations in pest management.

Collaboration and partnerships enhance the effectiveness of IPM. By working together, farmers, researchers, and extension agents can share information and develop region-specific strategies that address local pest challenges. Community-based IPM programs enable collective action, providing broader coverage and support for pest control efforts.

Policy and institutional support are crucial for advancing IPM adoption. Governments can promote IPM through incentives, research funding, and regulatory frameworks that encourage sustainable pest management practices. By providing access to resources and technical assistance, policymakers can help farmers transition to IPM and achieve long-term sustainability.

Consumer awareness and demand for sustainably produced food can drive the adoption of IPM practices. As consumers seek products that prioritize environmental and social responsibility, they indirectly support farmers who implement IPM. By choosing foods produced with sustainable pest management, consumers contribute to broader environmental and health benefits.

Biofertilizers and Natural Amendments

Biofertilizers and natural amendments are revolutionizing the way we approach soil fertility, offering sustainable alternatives to synthetic fertilizers while promoting healthy ecosystems. As the global agricultural community seeks methods that enhance productivity without compromising environmental integrity, these natural solutions are gaining traction. Their benefits extend beyond simply providing nutrients, as they play crucial roles in improving soil structure, promoting microbial activity, and enhancing plant resilience against pests and diseases.

Biofertilizers are comprised of living microorganisms, which, when applied to seeds, plant surfaces, or soil, colonize the rhizosphere or the interior of the plant and promote growth by increasing the supply or availability of primary nutrients. Among the most recognized biofertilizers are nitrogen-fixing bacteria, such as Rhizobium, which form symbiotic relationships with leguminous plants. These bacteria convert atmospheric nitrogen into a form that plants can use, significantly reducing the need for chemical nitrogen fertilizers. In return, the plant supplies the bacteria with carbohydrates, establishing a mutually beneficial relationship.

Phosphate-solubilizing bacteria and mycorrhizal fungi are other examples of biofertilizers that enhance nutrient availability. Phosphate-solubilizing bacteria release organic acids that convert insoluble phosphates in the soil into forms accessible to plants. Mycorrhizal fungi form associations with plant roots, extending their network through the soil and increasing the surface area for water and nutrient absorption. This symbiosis not only improves phosphorus uptake but also enhances the plant's ability to access other nutrients and water, particularly under stress conditions.

Natural amendments, such as compost, green manure, and vermicompost, enrich the soil by adding organic matter and nutrients that support plant growth. Composting involves the aerobic decomposition of organic materials, resulting in a nutrient-rich product that improves soil structure, water retention, and microbial diversity. By recycling plant and animal waste, composting reduces the need for chemical inputs and helps sequester carbon in the soil.

Green manure involves growing specific crops, such as legumes or cover crops, and then incorporating them into the soil to improve its fertility. These crops are typically grown during fallow periods and are turned under before they mature. As they decompose, they release nutrients back into the soil, enhance organic matter content, and improve soil tilth. Green manures also help suppress weeds and reduce erosion, contributing to overall farm sustainability.

Vermicompost is produced through the breakdown of organic material by earthworms. The resulting product is a nutrient-rich amendment that enhances soil fertility and supports a diverse microbial community. Vermicompost provides plants with essential nutrients, improves soil structure, and increases its water-holding capacity. Additionally, it has been shown to suppress plant diseases and improve plant growth, making it a valuable tool for sustainable agriculture.

The application of biofertilizers and natural amendments requires an understanding of soil conditions and crop needs. Soil testing is a vital step in determining the nutrient status and biological activity of the soil, guiding the choice and application rate of these inputs. Farmers must consider factors such as soil pH, texture, and organic matter content to tailor their soil management practices effectively.

One of the key advantages of using biofertilizers and natural amendments is their contribution to soil health and biodiversity. Unlike synthetic fertilizers, which can lead to nutrient imbalances and degrade soil structure, natural inputs support a thriving soil ecosystem. They promote the growth of beneficial microorganisms that play critical roles in nutrient cycling, disease suppression, and organic matter decomposition. Healthy soils, in turn, support robust plant growth and resilience to environmental stresses.

The integration of biofertilizers and natural amendments aligns with the principles of regenerative agriculture, which seeks to restore and enhance the natural functions of ecosystems. By reducing reliance on synthetic chemicals and focusing on soil health, farmers can create agricultural systems that are productive, resilient, and environmentally sustainable. This approach not only benefits the farm but also contributes to broader environmental goals, such as reducing greenhouse gas emissions and improving water quality.

Implementing these practices requires knowledge, experimentation, and adaptation. As with any agricultural technique, the effectiveness of biofertilizers and natural amendments depends on their proper use and management. Farmers are encouraged to start small, experiment with different combinations, and adjust their practices based on observations and results. Peer networks and extension services can provide valuable support and resources, fostering a community of learning and innovation.

Policy and institutional support are essential for promoting the adoption of biofertilizers and natural amendments. Governments can encourage their use through research funding, incentives, and technical assistance. By creating an enabling environment, policymakers can facilitate the transition to more sustainable agricultural practices and support farmers in achieving long-term productivity and environmental goals.

Consumer demand for sustainably produced food can further drive the adoption of these practices. As awareness of the environmental and health impacts of conventional agriculture grows, consumers increasingly seek products that prioritize sustainability. By supporting farmers who use biofertilizers and natural amendments, consumers contribute to the broader movement towards sustainable food systems and environmental stewardship.

Crop Rotation and Diversity

Crop rotation and diversity are foundational elements of sustainable agriculture, offering a suite of benefits that enhance soil health, reduce pest and disease pressures, and improve overall farm productivity. By strategically varying the types of crops grown in a particular area over time, farmers can harness the natural strengths of different plant species to create more resilient and productive agricultural systems.

The practice of crop rotation involves planting different types of crops in a sequential manner across seasons or years. This rotation disrupts the life cycles of pests and diseases that tend to specialize in particular crops, reducing their prevalence and impact. For example, a farmer might rotate nitrogen-fixing legumes with nutrient-demanding cereals. The legumes enrich the soil with nitrogen, benefiting the subsequent cereal crop, which in turn helps break the cycle of pests specific to legumes.

Crop diversity, on the other hand, refers to the simultaneous cultivation of a variety of crops within a single farm or field. This diversity can be achieved through intercropping, where different crops are grown together in the same space, or by maintaining a diverse mix of crops across different fields. Such diversity can enhance ecosystem services, such as pollination and pest control, by supporting a wide range of beneficial organisms.

One of the greatest benefits of crop rotation is its ability to improve soil fertility. Different crops have varying nutrient requirements and root structures, which influence how they interact with the soil. Deep-rooted crops can access nutrients from deeper soil layers, while shallow-rooted crops utilize nutrients nearer to the surface. By rotating crops with complementary nutrient needs and root profiles, farmers can maintain a balanced nutrient distribution in the soil, reducing the need for synthetic fertilizers.

In addition to nutrient management, crop rotation plays a crucial role in managing soil structure and organic matter content. Certain crops, like cover crops, are known for their ability to enhance soil structure by adding organic matter and improving soil porosity. When included in a rotation, these crops can improve water infiltration and retention, prevent erosion, and support a healthy soil microbiome. This not only enhances plant growth but also increases the soil's resilience to extreme weather conditions.

The impact of crop rotation on pest and disease management is profound. By regularly changing the plant environment, farmers can disrupt the life cycles of pests and pathogens, reducing their populations and minimizing the need for chemical pesticides. For instance, rotating a susceptible crop with a non-host crop can help break the cycle of soil-borne diseases. This natural form of pest and disease control is an integral component of integrated pest management strategies, promoting biodiversity and reducing chemical inputs.

Crop diversity also contributes to farm resilience by spreading risk. In a diverse cropping system, the failure of one crop due to pests, diseases, or adverse weather is less likely to lead to total crop failure, as other crops may still thrive. This diversity provides a buffer against market fluctuations and environmental stresses, contributing to the farm's economic stability and sustainability.

Incorporating crop rotation and diversity requires careful planning and knowledge of the local environment. Farmers must consider factors such as climate, soil type, and market demands when designing their cropping systems. Knowledge of plant families and their interactions is crucial for effective rotation planning. For example, rotating crops from different families can reduce the risk of disease carryover and nutrient depletion.

The implementation of crop rotation and diversity can be enhanced by using modern technologies and data management systems. Geographic information systems (GIS) and remote sensing can help farmers monitor field conditions and make informed decisions about crop placement and rotation schedules. By integrating traditional knowledge with modern technology, farmers can optimize their cropping systems for maximum productivity and sustainability.

Education and training are essential for promoting the adoption of crop rotation and diversity. Farmers and agricultural professionals need access to information and resources that enable them to implement these practices effectively. Workshops, extension services, and online resources can provide valuable guidance and support for designing and managing diverse cropping systems.

Policy and institutional support play a critical role in encouraging the use of crop rotation and diversity. Governments can promote these practices through incentives, research funding, and technical assistance. By creating an enabling environment, policymakers can help farmers transition to more sustainable agricultural practices and achieve long-term productivity and environmental goals.

Consumer demand for sustainably produced food can further drive the adoption of crop rotation and diversity. As awareness of the environmental and health impacts of conventional agriculture grows, consumers increasingly seek products that prioritize sustainability. By supporting farmers who use diverse cropping systems, consumers contribute to the broader movement towards sustainable food systems and environmental stewardship.

Genetically Modified Organisms: Pros and Cons

Genetically modified organisms (GMOs) represent a significant advancement in agricultural science, offering both opportunities and challenges. By altering the genetic makeup of crops and livestock, scientists aim to improve yield, resistance to pests and diseases, and adaptability to environmental stresses. As with any technological innovation, GMOs come with a set of pros and cons that must be carefully weighed by farmers, consumers, and policymakers.

On the positive side, GMOs have the potential to significantly increase agricultural productivity. Crops can be engineered to grow in less-than-ideal conditions, such as drought-prone or saline soils, thus expanding the areas available for cultivation. For example, drought-tolerant maize varieties have been developed to withstand periods of low rainfall, ensuring yield stability in regions where water scarcity is a growing concern. Similarly, salt-tolerant rice varieties enable farming in coastal areas affected by rising sea levels and soil salinity, a consequence of climate change.

Another advantage of GMOs is their contribution to pest and disease management. By incorporating genes that confer resistance to specific pests or diseases, genetically modified crops can reduce the need for chemical pesticides, lowering production costs and minimizing environmental impact. The Bt cotton, for instance, contains a gene from the bacterium Bacillus thuringiensis, which produces a protein toxic to certain insect pests. This built-in protection has led to a reduction in pesticide use and an increase in crop yields, benefiting both farmers and the environment.

GMOs also hold promise for enhancing the nutritional content of food. Biofortified crops, such as Golden Rice, have been engineered to contain higher levels of essential nutrients like vitamin A, addressing micronutrient deficiencies in populations that rely heavily on rice as a staple food. This nutritional enhancement can play a crucial role in improving public health, particularly in developing countries where access to diverse diets is limited.

Despite these benefits, GMOs are not without controversy. Concerns about the safety and long-term effects of genetically modified foods have been raised by consumers and advocacy groups. While numerous scientific studies have found GMOs to be safe for human consumption, skepticism remains, particularly around the potential for allergenicity or unintended health effects. Transparency and rigorous testing are essential to address these concerns and build public trust in GMO products.

Environmental implications are another area of concern. The widespread adoption of GMOs could lead to a reduction in biodiversity, as genetically modified varieties may outcompete traditional or wild relatives. This loss of genetic diversity can make ecosystems more vulnerable to pests, diseases, and climate change. Additionally, there is the risk of gene flow, where modified genes may transfer to non-target species, potentially creating "superweeds" or other unintended ecological consequences.

The socio-economic impacts of GMOs must also be considered. Smallholder farmers may face challenges accessing or affording genetically modified seeds, which are often patented and sold by large agribusinesses. This can lead to increased dependency on seed companies and reduce traditional farming practices, which rely on seed saving and exchange. Intellectual property rights and seed sovereignty are critical issues that need to be addressed to ensure equitable access to the benefits of GMOs.

Labeling and consumer choice are important aspects of the GMO debate. Many consumers desire the ability to make informed choices about the food they purchase and consume. Clear labeling of GMO products allows consumers to exercise this choice, respecting diverse values and preferences. Regulatory frameworks that mandate labeling and ensure the traceability of GMOs can help maintain transparency and consumer confidence.

Public perception and acceptance of GMOs vary widely across regions and cultures. In some countries, GMOs are embraced as a solution to food security challenges, while in others, they are met with resistance and regulatory hurdles. Education and communication play a vital role in bridging the gap between scientific advancements and public understanding. By fostering open dialogue and disseminating accurate information, stakeholders can build a more informed and balanced view of GMOs.

Policy and governance are crucial in shaping the development and deployment of GMOs. Governments must establish robust regulatory frameworks that ensure the safety and efficacy of genetically modified products while addressing ethical, environmental, and socio-economic considerations. Public-private partnerships and international cooperation can facilitate research, development, and dissemination of GMOs, ensuring that their benefits are accessible to all farmers, regardless of scale or location.

The future of GMOs lies in the ability to harness their potential responsibly and ethically. Continued research and innovation are essential for addressing the challenges of food security, climate change, and sustainable agriculture. By balancing the benefits and risks of GMOs, the agricultural community can develop strategies that maximize productivity while safeguarding human health and the environment.

Cover Cropping for Soil Enrichment

Cover cropping has long been heralded as a cornerstone of sustainable agriculture, offering numerous benefits that extend beyond the mere enrichment of soil. As an ancient practice, cover cropping involves the cultivation of specific plants, known as cover crops, primarily for the benefit of the soil rather than for harvest. These crops are typically grown during periods when main crops do not occupy the land, such as between growing seasons, and are either left to decompose on the surface or incorporated into the soil as green manure.

The choice of cover crop depends on the specific goals of the farmer, as different species offer unique advantages. Leguminous cover crops, like clover and vetch, are particularly prized for their ability to fix atmospheric nitrogen through symbiosis with soil bacteria. This natural nitrogen fixation enriches the soil, reducing the need for synthetic fertilizers in subsequent crops. As these legumes decompose, they release nitrogen slowly, ensuring a steady supply of this vital nutrient.

Grasses, such as rye and oats, are commonly used cover crops that excel at building soil organic matter and improving soil structure. Their extensive root systems help to capture and recycle nutrients that might otherwise leach away, while also enhancing soil aeration and water infiltration. As they break down, they contribute to the organic matter content, fostering a healthy soil microbiome and improving overall soil fertility.

Beyond nutrient management, cover crops are instrumental in soil conservation. Their root systems bind soil particles, reducing erosion caused by wind and water. This is particularly important in regions prone to heavy rainfall or strong winds, where topsoil loss can be significant. By protecting the soil surface, cover crops also mitigate the impact of raindrops, further preventing erosion and maintaining soil integrity.

Weed suppression is another key benefit of cover cropping. By occupying the soil during fallow periods, cover crops outcompete weeds for resources such as light, water, and nutrients. Some cover crops, like mustard and radish, even release natural biofumigants that suppress weed seed germination and growth. This natural form of weed control reduces the reliance on herbicides, contributing to more sustainable farming practices.

Cover crops also play a vital role in enhancing soil moisture retention. Their presence reduces evaporation from the soil surface, conserving water for the main crops. This is especially beneficial in arid regions or during periods of drought, where water scarcity poses a significant challenge to crop production. Additionally, the improved soil structure resulting from cover cropping enhances water infiltration and storage, further supporting plant growth during dry spells.

The integration of cover crops into farming systems requires careful planning and management. Farmers must select cover crops that align with their specific objectives, climate, and soil conditions. Timing is crucial, as cover crops need to be planted at the right time to maximize their benefits without interfering with the main crop's planting schedule. Similarly, the termination of cover crops—whether through mowing, rolling, or incorporation—must be well-timed to avoid competition with the main crops.

For beginners, implementing cover cropping might seem daunting, but the rewards are well worth the effort. Starting small, perhaps by trialing cover crops on a portion of the farm, allows for experimentation and learning. Farmers can observe the effects on soil health and crop productivity, adjusting their approach based on the outcomes. Local agricultural extension services and peer networks can provide valuable support and guidance, sharing knowledge and experiences that facilitate successful adoption.

Economic considerations are also important when adopting cover cropping. While there may be initial costs associated with purchasing seeds and additional labor, the long-term benefits often outweigh these expenses. Enhanced soil fertility can lead to reduced input costs, such as fertilizers and herbicides, while improvements in soil health and moisture retention can increase crop yields and resilience. Financial incentives or cost-sharing programs offered by government agencies or non-profit organizations can further support farmers in adopting cover cropping practices.

Cover cropping also aligns with broader environmental and sustainability goals. By enhancing soil health and reducing chemical inputs, cover crops contribute to improved water quality and reduced greenhouse gas emissions. The increase in soil organic matter sequesters carbon, playing a role in climate change mitigation. Additionally, cover crops support biodiversity, providing habitat and food sources for beneficial insects and wildlife.

The cultural and societal benefits of cover cropping should not be overlooked. By fostering sustainable land management practices, cover cropping supports rural communities and promotes food security. Farmers who adopt these practices often become stewards of the land, contributing to the long-term health and productivity of their local ecosystems.

Chapter 4

Embracing AgriTech

Smart Farming Technologies

Smart farming technologies are transforming agriculture, ushering in an era where data-driven insights and automation enhance productivity, sustainability, and resource efficiency. By leveraging advanced tools and techniques, farmers can make more informed decisions, optimize their operations, and address the challenges posed by climate change, population growth, and resource scarcity.

Precision agriculture is at the heart of smart farming technologies. It involves the use of GPS, remote sensing, and data analytics to monitor and manage crop production at a granular level. By collecting data on soil conditions, weather patterns, and crop health, farmers can tailor their practices to the specific needs of each field or even individual plants. This targeted approach reduces waste, improves yields, and minimizes environmental impact.

Drones have become invaluable tools in modern agriculture, providing farmers with real-time aerial imagery and data. Equipped with multispectral and thermal cameras, drones can assess crop health, detect pest infestations, and monitor water stress. This information enables farmers to implement timely interventions, such as applying pesticides or adjusting irrigation, ensuring optimal crop performance and resource use.

The Internet of Things (IoT) is another key component of smart farming, connecting devices and sensors across the farm to create an integrated network. IoT-enabled sensors can monitor soil moisture, temperature, and nutrient levels, providing farmers with continuous updates on field conditions. This real-time data allows for precise irrigation and fertilization, reducing water and chemical usage while maximizing crop growth.

Autonomous machinery, such as self-driving tractors and robotic harvesters, is revolutionizing farm labor. These machines can perform tasks with precision and efficiency, reducing the need for manual labor and increasing productivity. By automating repetitive and labor-intensive processes, farmers can focus on higher-level decision-making and management tasks, improving overall operational efficiency.

Big data analytics and artificial intelligence (AI) play a crucial role in smart farming by processing vast amounts of information to generate actionable insights. Advanced algorithms can analyze historical and real-time data to predict crop yields, identify optimal planting times, and forecast pest outbreaks.

By leveraging these insights, farmers can make proactive decisions that enhance productivity and minimize risks.

Blockchain technology offers transparency and traceability in agriculture, addressing issues related to food safety, quality control, and supply chain management. By recording transactions and data on a decentralized ledger, blockchain ensures that information is secure and tamper-proof. This technology can help farmers verify the authenticity of their products, meet regulatory requirements, and build trust with consumers.

The benefits of smart farming technologies extend beyond productivity gains. By improving resource efficiency and reducing waste, these technologies contribute to environmental sustainability. Precision agriculture, for example, minimizes the over-application of fertilizers and pesticides, reducing nutrient runoff and pollution. Similarly, optimized irrigation practices conserve water resources and mitigate the impact of drought.

Implementing smart farming technologies requires investment and technical expertise. Farmers must assess their specific needs and capabilities to determine which technologies are most suitable for their operations. Collaboration with technology providers, agricultural consultants, and research institutions can provide valuable support and guidance during the adoption process.

Education and training are essential to ensure that farmers can effectively utilize smart farming technologies. Workshops, online courses, and peer networks can provide the knowledge and skills needed to operate and maintain advanced tools and systems. By fostering a culture of continuous learning and innovation, the agricultural community can stay at the forefront of technological advancements.

Policy and institutional support are crucial for promoting the adoption of smart farming technologies. Governments can facilitate access to technology through funding, incentives, and infrastructure development. By creating an enabling environment, policymakers can help farmers transition to more efficient and sustainable practices, ensuring long-term agricultural productivity and resilience.

Consumer demand for transparency and sustainability drives the adoption of smart farming technologies. As consumers seek to know more about the origin and production methods of their food, farmers can use technology to provide verifiable information and meet these expectations. By aligning production practices with consumer values, farmers can enhance market access and competitiveness.

The integration of smart farming technologies presents opportunities and challenges. While these technologies offer significant benefits, they also raise concerns related to data privacy, security, and equity. Ensuring that all farmers, regardless of size or location, have access to technology and the resources needed to implement it is critical for equitable development.

Drones and Satellite Imagery

Drones and satellite imagery are revolutionizing the agricultural landscape, offering innovative solutions for monitoring, analysis, and management of crops. These technologies provide farmers with unprecedented access to data and insights that drive precision agriculture, enhancing productivity and sustainability.

The use of drones in agriculture has surged in recent years due to their versatility and efficiency. Equipped with advanced cameras and sensors, drones fly over fields to capture high-resolution images and data. This aerial perspective allows farmers to assess crop health, identify pest infestations, and monitor water stress with remarkable accuracy. The ability to quickly gather detailed information over large areas is invaluable, enabling timely interventions that can prevent minor issues from escalating into major problems.

A key advantage of drones lies in their capacity to provide multispectral and thermal imagery. Multispectral cameras capture images in different wavelengths of light, revealing information that is invisible to the naked eye. This data can be used to calculate vegetation indices, such as the Normalized Difference Vegetation Index (NDVI), which indicates plant health and vigor. Thermal cameras, on the other hand, detect temperature variations, helping farmers identify areas of heat stress or uneven irrigation. By integrating these insights, farmers can make informed decisions to optimize crop management.

In addition to drones, satellite imagery plays a pivotal role in modern agriculture. Satellites orbiting the Earth continuously capture images of the planet's surface, providing a comprehensive view of agricultural landscapes. These images are invaluable for tracking changes over time, such as crop growth patterns, land use, and soil conditions. Satellite data is particularly useful for large-scale farms or regions where drone deployment may be challenging.

One of the primary benefits of satellite imagery is its ability to cover vast areas, offering a macro-level perspective that complements the detailed, localized data provided by drones. This combination of micro and macro insights empowers farmers to understand their fields at multiple scales, facilitating more effective planning and resource allocation. For instance, satellite imagery can help identify areas that require further investigation with drones or ground-based inspections, streamlining the decision-making process.

The integration of drones and satellite imagery into precision agriculture is facilitated by advanced data processing and analytics. Software platforms analyze the collected data, generating actionable insights that guide farm management. These platforms often employ machine learning algorithms to detect patterns and anomalies, predict crop yields, and recommend optimal practices. By leveraging these technologies, farmers can enhance productivity while minimizing inputs and environmental impact.

One practical application of drones and satellite imagery is in precision irrigation. By analyzing data on soil moisture and crop water needs, farmers can adjust irrigation schedules and amounts to match actual requirements. This targeted approach conserves water, reduces costs, and prevents issues related to overwatering or underwatering. In regions where water scarcity is a concern, precision irrigation can significantly improve sustainability and resilience.

Drones and satellite imagery also contribute to precision fertilization. By mapping nutrient levels across fields, farmers can apply fertilizers more efficiently, ensuring that each area receives the appropriate amount. This precision reduces waste and environmental impact while promoting uniform crop growth and maximizing yields. Furthermore, by monitoring crop health over time, farmers can track the effectiveness of fertilization strategies and adjust them as needed.

The use of drones and satellite imagery extends beyond individual farm management, offering valuable insights for regional and global agricultural monitoring. Governments and organizations can use these technologies to assess food security, monitor climate change impacts, and plan for disaster response. Satellite data, for example, can track changes in agricultural land use, alerting authorities to deforestation or land degradation. This information is crucial for developing policies and practices that promote sustainable land management.

While the benefits of drones and satellite imagery are clear, their adoption requires investment and expertise. Farmers must evaluate their specific needs and resources to determine which technologies are most appropriate. Initial costs for equipment, software, and training can be substantial, but the long-term benefits often outweigh these expenses.

Collaborative efforts with technology providers, research institutions, and agricultural consultants can support the successful integration of these tools into farming operations.

Education and training are essential for farmers to effectively use drones and satellite imagery. Workshops, online courses, and peer networks can provide the knowledge and skills needed to operate these technologies and interpret the data they generate. By fostering a culture of innovation and continuous learning, the agricultural community can maximize the potential of these advanced tools.

Policy and infrastructure support are also crucial for promoting the adoption of drones and satellite imagery. Governments can facilitate access to technology through funding, subsidies, and regulatory frameworks that ensure data privacy and security. By creating an enabling environment, policymakers can help farmers transition to more efficient and sustainable practices, ensuring long-term agricultural productivity and resilience.

IoT in Agriculture

The Internet of Things (IoT) is reshaping the agricultural landscape, bringing about a new era of connectivity and data-driven decision-making. By linking various devices and systems, IoT enables farmers to monitor and manage agricultural operations with unprecedented precision and efficiency. This technological advancement offers transformative solutions to some of the most pressing challenges in modern agriculture, such as resource management, productivity enhancement, and sustainability.

At its core, IoT in agriculture involves the use of interconnected sensors and devices that collect and transmit data about various aspects of the farming environment. These IoT-enabled sensors can be placed throughout the farm to measure soil moisture, temperature, humidity, light levels, and other critical parameters. This constant stream of data provides farmers with real-time insights into the conditions of their fields, enabling them to make informed decisions that optimize crop growth and resource use.

One of the most significant applications of IoT in agriculture is precision irrigation. By using soil moisture sensors connected to an IoT network, farmers can determine the exact water needs of their crops. This information allows for the precise application of water, reducing waste and conserving this vital resource. In regions prone to drought or water scarcity, precision irrigation can significantly enhance sustainability and crop resilience, ensuring that water is applied only where and when it is needed.

IoT also plays a crucial role in precision fertilization. Nutrient sensors placed in the soil can detect nutrient levels and plant needs, allowing for targeted fertilizer application. By delivering fertilizers precisely where they are required, farmers can reduce chemical runoff, minimize environmental impact, and promote healthier plant growth. This approach not only benefits the environment but also lowers input costs and maximizes yield potential.

Crop monitoring is another area where IoT offers substantial advantages. IoT devices can track plant health, growth rates, and pest activity, providing farmers with timely alerts and actionable data. For example, sensors can detect early signs of pest infestations or disease outbreaks, enabling farmers to intervene before the problem becomes widespread. This proactive approach reduces the reliance on chemical pesticides and enhances overall crop health and quality.

Beyond field management, IoT extends to livestock farming as well. Wearable sensors for animals can monitor vital signs, movement, and behavior, offering insights into health and well-being. Farmers can receive alerts about abnormal conditions, such as illness or injury, allowing for prompt intervention and care. This technology not only improves animal welfare but also enhances productivity by ensuring that livestock remains healthy and stress-free.

IoT in agriculture is supported by sophisticated data analytics platforms that process and interpret the vast amounts of data generated by sensors. These platforms use machine learning algorithms to identify patterns, predict outcomes, and recommend optimal practices. By providing farmers with actionable insights, data analytics enhance decision-making and operational efficiency, leading to increased productivity and reduced risk.

The implementation of IoT in agriculture requires careful planning and investment. Farmers must evaluate their specific needs and capabilities to select the appropriate technologies and devices. Initial costs can be significant, including the purchase of sensors, connectivity infrastructure, and data management systems. However, the long-term benefits of increased efficiency, reduced resource use, and improved yields often justify these investments.

Education and training are essential for farmers to effectively utilize IoT technologies. Understanding how to install, operate, and maintain IoT devices is crucial for maximizing their potential. Workshops, training programs, and support from technology providers can equip farmers with the necessary skills and knowledge to leverage IoT effectively. By fostering a culture of innovation and continuous learning, the agricultural community can remain at the forefront of technological advancements.

Policy and infrastructure support are also critical for the widespread adoption of IoT in agriculture. Governments can facilitate access to technology through funding, subsidies, and the development of rural connectivity infrastructure. By creating an enabling environment, policymakers can help farmers transition to more efficient and sustainable practices, ensuring long-term agricultural productivity and resilience.

IoT technology brings transparency and traceability to the agricultural supply chain, addressing consumer demand for information about food origin and production practices. By providing detailed data about cultivation, harvest, and processing, IoT enables farmers to meet regulatory requirements and build trust with consumers. This transparency enhances market access and competitiveness, aligning production practices with consumer values.

The integration of IoT in agriculture is not without challenges. Issues related to data security, privacy, and interoperability must be addressed to ensure the safe and efficient use of IoT technologies. Farmers need assurance that their data is protected and that IoT devices are compatible with existing systems and infrastructure. Collaborative efforts among technology providers, policymakers, and agricultural stakeholders are necessary to tackle these challenges and drive progress.

The future of IoT in agriculture is promising, with continued advancements in sensor technology, data analytics, and connectivity expected to further enhance the capabilities and impact of IoT solutions. As the global population grows and environmental pressures intensify, the need for efficient and sustainable agricultural practices becomes ever more critical. IoT technology offers a pathway to meet these challenges, empowering farmers to optimize their operations and contribute to a more secure and resilient food system.

Robotics and Automation

Robotics and automation are reshaping the agricultural industry, offering transformative solutions that enhance efficiency, productivity, and sustainability. By automating labor-intensive tasks and integrating advanced robotic systems, farmers can optimize their operations, reduce costs, and address the challenges of labor shortages and increasing demand for food.

The introduction of robotic technologies in agriculture has been driven by the need to improve efficiency and precision in farming practices. These technologies range from autonomous tractors and robotic harvesters to drones and automated milking systems. Each of these innovations brings unique benefits, contributing to the overall advancement of agricultural practices.

Autonomous tractors are at the forefront of agricultural robotics, capable of performing a wide range of tasks without human intervention. These tractors are equipped with GPS and advanced sensors that allow them to navigate fields with precision, perform tasks such as plowing, planting, and spraying, and adapt to varying field conditions. By automating these repetitive tasks, farmers can save time, reduce labor costs, and improve the overall efficiency of their operations.

Robotic harvesters are another significant advancement in agricultural automation. These machines are designed to pick fruits and vegetables with precision, minimizing damage and ensuring optimal quality. Equipped with cameras and sensors, robotic harvesters can identify ripe produce and selectively harvest it, reducing waste and improving yield. This technology is particularly beneficial in labor-intensive crops such as strawberries, apples, and grapes, where manual harvesting can be time-consuming and costly.

In dairy farming, robotic milking systems have revolutionized the milking process, offering increased efficiency and improved animal welfare. These systems allow cows to be milked at their convenience, reducing stress and increasing milk yield. The robotic milking process is highly automated, with sensors and software managing the milking routine, monitoring milk quality, and keeping track of each cow's health. This technology not only enhances productivity but also frees up farmers to focus on other management tasks.

Automation extends beyond fieldwork, with advancements in processing and packaging also contributing to the efficiency of agricultural operations. Automated sorting and grading systems use cameras and sensors to assess crop quality, ensuring that only the best produce reaches the market. These systems can sort and package produce quickly and accurately, reducing labor costs and minimizing human error.

The integration of robotics and automation in agriculture is supported by advanced data analytics and machine learning. These technologies enable robots to learn and adapt to varying conditions, improving their performance and decision-making capabilities. By analyzing data collected from sensors and cameras, robotic systems can optimize their operations, ensuring that tasks are carried out with precision and efficiency.

The adoption of robotics and automation in agriculture requires investment and careful planning. Farmers must assess their specific needs and capabilities to select the appropriate technologies and systems. Initial costs for purchasing and implementing robotic solutions can be substantial, but the long-term benefits often justify these expenses. Increased efficiency, reduced labor costs, and improved yields contribute to a positive return on investment for farmers.

Education and training are essential for farmers to effectively utilize robotic technologies. Understanding how to operate and maintain these systems is crucial for maximizing their potential. Training programs, workshops, and support from technology providers can equip farmers with the necessary skills and knowledge to leverage robotics effectively. By fostering a culture of innovation and continuous learning, the agricultural community can remain at the forefront of technological advancements.

Policy and infrastructure support are also critical for promoting the adoption of robotics and automation in agriculture. Governments can facilitate access to technology through funding, incentives, and the development of rural connectivity infrastructure. By creating an enabling environment, policymakers can help farmers transition to more efficient and sustainable practices, ensuring long-term agricultural productivity and resilience.

The integration of robotics and automation into agriculture offers significant benefits but also raises challenges related to data security, privacy, and workforce displacement. Ensuring that data collected by robotic systems is protected and that workers are reskilled for new roles in automated environments is crucial for the successful adoption of these technologies. Collaborative efforts among technology providers, policymakers, and agricultural stakeholders are necessary to address these challenges and drive progress.

The future of robotics and automation in agriculture is promising, with continued advancements in sensor technology, data analytics, and connectivity expected to further enhance the capabilities and impact of robotic solutions. As the global population grows and environmental pressures intensify, the need for efficient and sustainable agricultural practices becomes ever more critical. Robotics and automation offer a pathway to meet these challenges, empowering farmers to optimize their operations and contribute to a more secure and resilient food system.

Blockchain for Agricultural Supply Chains

Blockchain technology is increasingly finding its place in agricultural supply chains, offering a robust solution to many of the challenges faced by the sector. Its decentralized and transparent nature ensures that every transaction and data point is recorded immutably, providing a new level of trust and efficiency in the way food is produced, transported, and consumed.

Understanding the complexities of agricultural supply chains reveals why blockchain is so transformative. These supply chains involve multiple stakeholders, including farmers, processors, distributors, retailers, and consumers. Each step involves the transfer of goods and information, which, when managed inefficiently, can lead to delays, errors, and even fraud. Blockchain brings clarity and accountability to these processes, ensuring that every participant in the chain is on the same page.

One of the primary benefits of blockchain in agriculture is the enhancement of traceability. Consumers increasingly demand transparency regarding the origins of their food, seeking assurance that products are sustainably sourced and ethically produced. Blockchain provides an unalterable record of a product's journey from farm to table, capturing data such as the location of production, handling processes, and delivery timelines. This traceability not only satisfies consumer demands but also helps producers and retailers quickly identify and address issues, such as contamination or recalls, that could otherwise lead to significant economic losses and reputational damage.

The integrity of data is another crucial aspect where blockchain excels. In traditional supply chains, data can be altered, tampered with, or lost, leading to mistrust among stakeholders. Blockchain's decentralized ledger ensures that once data is recorded, it cannot be changed. This feature is particularly valuable for verifying the authenticity of organic or fair-trade claims, as stakeholders can trust that the information has not been manipulated at any point in the process.

Blockchain also facilitates smart contracts—self-executing contracts with the terms directly written into code. These contracts automatically enforce and execute agreements when predefined conditions are met, reducing the need for intermediaries and minimizing disputes. For farmers, this can mean faster payments and reduced administrative burdens, as payments can be automatically released upon the receipt of goods or services that meet agreed-upon standards.

The potential of blockchain extends to improving efficiency and reducing costs across agricultural supply chains. By providing a single, shared ledger accessible to all participants, blockchain reduces the need for repetitive data entry and verification, streamlining operations and cutting down on paperwork. This efficiency can lead to cost savings, which are particularly beneficial for smallholder farmers and producers who often operate on tight margins.

Adopting blockchain technology in agriculture requires a collaborative approach. Stakeholders across the supply chain must be willing to share data and trust the system. This shift may require a cultural change, as well as investment in technology and training. Farmers and other participants need to understand how to input data accurately and how to access and interpret the information stored on the blockchain.

The implementation of blockchain in agricultural supply chains does face challenges, such as the digital divide that exists in many rural areas. Access to reliable internet and digital literacy are prerequisites for participating in blockchain-based systems. Efforts must be made to bridge this gap, ensuring that all stakeholders, especially those in developing regions, can benefit from the technology.

Policy support and regulatory frameworks are essential to facilitate the adoption of blockchain in agriculture. Governments can play a crucial role by providing incentives, developing standards, and ensuring that blockchain implementations comply with existing laws and regulations. This support can help build trust in the technology and encourage its widespread use.

The environmental impact of blockchain technology is another consideration. While blockchain offers significant benefits, it can also be energy-intensive, particularly public blockchains that rely on proof-of-work consensus mechanisms. Exploring more sustainable consensus methods, such as proof-of-stake or consortium blockchains, can help mitigate this impact and align with the agricultural sector's sustainability goals.

Blockchain has the potential to transform not only agricultural supply chains but also the broader food system. By ensuring transparency, enhancing trust, and improving efficiency, blockchain can contribute to a more sustainable and equitable food future. As the technology continues to evolve, its integration into agriculture will likely expand, offering new opportunities for innovation and collaboration.

Chapter 5

Sustainable Livestock Practices

Ethical Animal Husbandry

Ethical animal husbandry is gaining recognition as an essential component of sustainable agriculture, reflecting a growing awareness of the need to treat animals with respect and care throughout their life cycles. It encompasses practices that prioritize the welfare of livestock, ensuring their physical and mental well-being while maintaining productivity and efficiency. In an era where consumers are increasingly concerned about the ethical implications of their food choices, adopting humane and responsible animal husbandry practices is not only a moral obligation but also a strategic advantage for farmers.

The foundation of ethical animal husbandry lies in understanding the natural behaviors and needs of different livestock species. Each species has unique requirements for space, nutrition, social interaction, and environmental enrichment. By recognizing and accommodating these needs, farmers can create an environment that promotes health, reduces stress, and enhances overall productivity. For example, providing ample space for movement and social interaction is crucial for herd animals such as cattle and sheep, while chickens benefit from access to outdoor areas where they can forage and dust-bathe.

Nutrition plays a critical role in ethical animal husbandry, as a balanced diet is essential for the health and growth of livestock. Farmers must ensure that animals receive adequate nutrients tailored to their specific needs, taking into account factors such as age, breed, and production stage. Providing high-quality feed, clean water, and appropriate supplementation not only supports animal health but also improves the quality of the products they produce, such as milk, eggs, and meat.

Housing and shelter are also vital components of ethical animal husbandry. Livestock should be provided with clean, comfortable living conditions that protect them from extreme weather conditions and allow for natural behaviors. Proper ventilation, temperature control, and sanitation are essential to prevent the spread of diseases and maintain a healthy environment. For instance, dairy cows housed in well-ventilated barns with clean bedding are less prone to respiratory issues and infections, leading to better milk production and overall health.

Preventive healthcare is a cornerstone of ethical animal husbandry, emphasizing the importance of disease prevention and early intervention. Regular veterinary check-ups, vaccinations, and parasite control are essential to safeguard animal health and prevent outbreaks of contagious diseases. Farmers should also be trained to recognize early signs of illness or distress, enabling them to take prompt action and minimize suffering. By prioritizing preventive care, farmers can reduce the need for antibiotics and other medications, contributing to the fight against antibiotic resistance.

In addition to physical well-being, ethical animal husbandry considers the mental health and welfare of livestock. Providing environmental enrichment, such as toys, obstacles, and varied terrains, can prevent boredom and promote natural behaviors, reducing stress and improving overall well-being. Social animals, like pigs and chickens, thrive in environments that allow them to interact with their peers, while solitary animals, like some breeds of goats, may require individual space to prevent stress and aggression.

Handling and transport are critical aspects of ethical animal husbandry that require careful consideration and planning. Gentle handling techniques, such as using voice commands and minimizing physical contact, reduce stress and prevent injury during routine procedures like vaccination, shearing, or milking. Transporting animals should be done with care, ensuring that vehicles are well-ventilated, non-slip, and spacious enough to allow animals to stand and lie down comfortably. Minimizing transport time and providing rest breaks can further reduce stress and improve welfare during transit.

Ethical animal husbandry also involves responsible breeding practices that prioritize the health and welfare of both parent animals and offspring. Selective breeding should focus on traits that enhance animal well-being, such as disease resistance, temperament, and adaptability to local conditions, rather than solely on productivity. Avoiding inbreeding and ensuring genetic diversity within herds or flocks can prevent health issues and improve resilience to environmental challenges.

Education and training are essential for farmers to implement ethical animal husbandry practices effectively. Access to resources, workshops, and support from agricultural organizations can equip farmers with the knowledge and skills needed to enhance animal welfare on their farms. By fostering a culture of continuous learning and improvement, the agricultural community can advance towards more humane and sustainable practices.

Consumer demand for ethically produced animal products is a driving force behind the adoption of ethical animal husbandry. As consumers become more informed about the origins of their food, they seek products that align with their values, such as animal welfare certifications or labels indicating humane treatment. By meeting these expectations, farmers can enhance their market access and competitiveness, building trust with consumers and contributing to a more sustainable food system.

Policy support and regulatory frameworks are crucial for promoting ethical animal husbandry practices. Governments can play a significant role by establishing welfare standards, providing incentives for humane farming practices, and supporting research into innovative husbandry techniques. By creating an enabling environment, policymakers can help farmers transition to more ethical practices, ensuring long-term agricultural sustainability and resilience.

Pasture-Based Systems

Pasture-based systems hold a unique place in the realm of agriculture, offering a harmonious blend of sustainability, animal welfare, and resource efficiency. This approach to farming emphasizes the use of natural landscapes to raise livestock, allowing animals to graze on pastures and benefit from a diet that closely mimics their natural feeding behaviors. By prioritizing the health of both the land and the animals, pasture-based systems present a compelling alternative to more intensive farming methods.

The allure of pasture-based systems lies in their simplicity and alignment with ecological principles. Grazing animals on pastures can improve soil health, increase biodiversity, and reduce the reliance on external inputs such as fertilizers and feed supplements. As herbivores graze, they naturally fertilize the land with their manure, contributing to a nutrient-rich soil ecosystem that supports plant growth. This cycle of grazing and regrowth promotes a resilient and self-sustaining agricultural system.

One of the most significant advantages of pasture-based systems is the potential for enhanced animal welfare. Livestock that are allowed to graze freely on pastures experience a more natural and less stressful environment compared to those raised in confined spaces. The ability to roam, forage, and engage in natural behaviors can lead to improved physical and mental well-being, resulting in healthier and more productive animals. For example, cows grazing on pasture tend to have lower stress levels, which can positively impact milk yield and quality.

Pasture-based systems also offer environmental benefits by sequestering carbon in the soil. As grasses and plants grow, they capture carbon dioxide from the atmosphere and store it in their roots and surrounding soil. This process not only helps mitigate climate change but also enhances soil structure and fertility. Well-managed pastures can act as carbon sinks, offsetting greenhouse gas emissions and contributing to a more sustainable agricultural landscape.

For farmers, transitioning to pasture-based systems can lead to cost savings and increased resilience. By reducing dependence on purchased feeds and synthetic inputs, farmers can lower production costs and improve their bottom line. Additionally, pasture-based systems are often more resilient to market fluctuations and supply chain disruptions, as they rely on homegrown resources and local ecosystems. This self-sufficiency can be particularly valuable in times of economic uncertainty or environmental challenges.

Implementing a successful pasture-based system requires careful planning and management. Farmers must assess their land's capacity, taking into account factors such as soil type, climate, and water availability. Rotational grazing, where livestock are moved between different pasture areas, is a key strategy for maintaining pasture health and preventing overgrazing. By allowing pastures to rest and regenerate, farmers can ensure a continuous supply of nutritious forage for their animals.

Water management is another critical aspect of pasture-based systems. Access to clean and reliable water sources is essential for animal health and productivity. Farmers may need to invest in water infrastructure, such as ponds, wells, or irrigation systems, to ensure that grazing animals have access to adequate hydration. Efficient water use not only supports livestock welfare but also contributes to the overall sustainability of the farming operation.

Pasture-based systems can also enhance biodiversity by providing habitats for a variety of plant and animal species. Diverse pastures with a mix of grasses, legumes, and herbs can support pollinators, beneficial insects, and wildlife, contributing to a balanced and thriving ecosystem. By fostering biodiversity, farmers can benefit from natural pest control and pollination services, reducing the need for chemical inputs and enhancing the resilience of their farming systems.

Education and training are essential for farmers considering a transition to pasture-based systems. Access to resources, workshops, and mentorship can equip farmers with the knowledge and skills needed to manage pastures effectively and optimize animal health and productivity. By fostering a culture of learning and collaboration, the agricultural community can support the widespread adoption of pasture-based practices.

Consumer demand for sustainably and ethically produced food is a driving force behind the growing interest in pasture-based systems. As consumers become more conscious of the environmental and ethical implications of their food choices, they increasingly seek products that align with their values, such as grass-fed beef or pasture-raised eggs. By meeting these expectations, farmers can enhance their market access and competitiveness, building trust with consumers and contributing to a more sustainable food system.

Policy support and regulatory frameworks are crucial for promoting pasture-based systems. Governments can play a significant role by providing incentives, developing standards, and supporting research into innovative grazing techniques. By creating an enabling environment, policymakers can help farmers transition to more sustainable practices, ensuring long-term agricultural productivity and resilience.

Alternative Feed Sources

The search for alternative feed sources in agriculture is driven by the need to improve sustainability, reduce costs, and enhance the nutritional quality of livestock diets. Traditional feed ingredients, like corn and soybean meal, often come with environmental

and economic challenges, including resource-
intensive production and market volatility.
Diversifying feed sources can mitigate these issues
while promoting more resilient and sustainable
farming practices.

One promising avenue for alternative feed is the
incorporation of insect-based proteins. Insects, such
as black soldier fly larvae and mealworms, offer a
high-protein, nutrient-dense option that can be
produced sustainably. These larvae can be raised on
organic waste, converting it into valuable protein and
reducing the environmental footprint of feed
production. Insect farming requires significantly less
land, water, and resources compared to conventional
livestock feed, making it an attractive option for eco-
conscious farmers.

Seaweed is another alternative feed source gaining
attention for its nutritional benefits and
environmental advantages. Rich in vitamins,
minerals, and antioxidants, seaweed can enhance
livestock health and productivity when included in
their diets. Moreover, certain types of seaweed have
been shown to reduce methane emissions in
ruminants, contributing to climate change mitigation
efforts. As seaweed farming expands, it promises a
renewable and sustainable feed option that can
support both land-based and aquatic farming
systems.

Agricultural by-products, often considered waste, can also serve as valuable feed components. Items such as brewers' grains, citrus pulp, and beet pulp are rich in nutrients and can be incorporated into livestock diets, reducing the need for traditional feed ingredients. Utilizing these by-products not only decreases feed costs but also minimizes waste, aligning with circular economy principles. Farmers can work with local industries to source these materials, fostering regional collaboration and sustainability.

Cover crops, traditionally used to improve soil health and reduce erosion, can also be leveraged as alternative feed. Species such as clover, alfalfa, and rye can be grown as dual-purpose crops, providing both soil benefits and livestock nutrition. By integrating cover crops into crop rotations, farmers can increase feed availability while enhancing soil fertility and biodiversity. This practice exemplifies a holistic approach to farming that balances productivity with ecological stewardship.

Exploring alternative grains and legumes is another strategy to diversify livestock diets. Grains like barley, oats, and sorghum, along with legumes such as lupins and peas, offer nutritional profiles that can complement or replace conventional feed components. These crops often require fewer inputs and are more resilient to climate variability, making them suitable for sustainable agriculture systems. By incorporating a variety of grains and legumes, farmers can reduce dependency on monocultures and enhance the resilience of their feed supply.

The development and adoption of alternative feed sources require research, innovation, and collaboration across the agricultural sector. Scientists and agronomists play a crucial role in evaluating the nutritional value, safety, and environmental impact of new feed options. Collaborative research efforts can lead to breakthroughs in feed formulation and processing, ensuring that alternative sources meet the dietary needs of livestock while supporting sustainable practices.

Farmers interested in adopting alternative feed sources must consider the logistics of sourcing, processing, and integrating these ingredients into their operations. Building relationships with suppliers, investing in appropriate storage and processing equipment, and adjusting feeding strategies are essential steps in this transition. Education and training programs can provide farmers with the knowledge and skills needed to successfully implement alternative feed strategies, ensuring that livestock health and productivity are maintained.

Policy support and incentives can accelerate the adoption of alternative feed sources by reducing barriers and encouraging innovation. Governments can play a pivotal role by funding research, establishing quality standards, and facilitating market access for alternative feed ingredients. By creating a supportive regulatory environment, policymakers can help farmers transition to more sustainable and resilient feed systems.

Consumer demand for sustainably produced animal products can drive the expansion of alternative feed sources. As awareness of environmental and ethical issues grows, consumers increasingly seek products from farms that prioritize sustainability and innovation. By meeting these expectations, farmers can enhance their market competitiveness and build trust with consumers, contributing to a more sustainable food system.

The exploration of alternative feed sources represents a significant opportunity for agriculture to address pressing environmental and economic challenges. By diversifying feed ingredients, farmers can reduce their reliance on resource-intensive crops, decrease environmental impact, and enhance the resilience of their operations. Through collaboration, research, and a commitment to sustainability, the agricultural community can unlock the potential of alternative feed sources, paving the way for a more sustainable and resilient future.

Waste Management and Recycling

Efficient waste management and recycling practices are essential components of modern agriculture, addressing both environmental concerns and economic opportunities. As agriculture produces significant amounts of waste, ranging from crop residues to livestock manure, developing strategies to manage and repurpose these materials is crucial for sustainable farming. By transforming waste into valuable resources, farmers can reduce environmental impact, improve soil health, and enhance farm profitability.

Organic waste, including crop residues, manure, and food by-products, forms a substantial part of agricultural waste. These materials, if not managed properly, can contribute to greenhouse gas emissions, water contamination, and nutrient runoff. However, when effectively recycled, they serve as vital resources for improving soil fertility and structure. Composting is one of the most accessible and beneficial methods of recycling organic waste. By decomposing organic matter into nutrient-rich compost, farmers can create a natural fertilizer that enhances soil health and reduces the need for chemical inputs. Composting not only recycles nutrients but also improves soil water retention and supports microbial activity, fostering a vibrant and productive ecosystem.

Livestock manure, often considered a waste product, can be a valuable asset when managed correctly. Animal waste contains essential nutrients such as nitrogen, phosphorus, and potassium, which are vital for plant growth. By applying manure to fields as a natural fertilizer, farmers can enrich their soils organically. Proper manure management, including storage and application techniques, is critical to prevent nutrient leaching and emissions. Technologies such as anaerobic digestion can also convert manure into biogas, providing renewable energy for farm operations and reducing reliance on fossil fuels.

Crop residues, often left in fields after harvest, present another opportunity for recycling. These materials can be incorporated into the soil to improve organic matter content, reduce erosion, and enhance soil structure. Alternatively, they can be used as feedstock for bioenergy production, providing an additional income stream for farmers. The key to successful residue management is balancing the benefits of soil incorporation with the potential uses for bioenergy or animal feed.

Plastic waste, particularly from packaging and agricultural films, poses a unique challenge in agriculture. Addressing this issue requires a multifaceted approach, including reducing plastic use, promoting biodegradable alternatives, and improving recycling systems. Farmers can adopt practices such as using reusable containers, selecting biodegradable mulches, and participating in recycling programs to minimize plastic waste. Collaborating with suppliers and policymakers to develop sustainable packaging solutions can also contribute to reducing the environmental footprint of agriculture.

Water management is intrinsically linked to waste management in agriculture. Efficient irrigation systems and water recycling practices can reduce water waste and conserve valuable resources. Technologies such as drip irrigation and rainwater harvesting enable farmers to use water more efficiently, while constructed wetlands and bioreactors can treat agricultural runoff, preventing pollution and protecting water quality. By integrating water-saving practices into their operations, farmers can enhance sustainability and resilience in the face of climate change.

Education and training are vital for successful waste management and recycling in agriculture. Farmers need access to information and resources to implement effective strategies, from composting techniques to energy recovery systems. Workshops, extension services, and peer networks can provide valuable support and foster a culture of innovation and continuous improvement. By sharing knowledge and experiences, the agricultural community can collectively advance toward more sustainable waste management practices.

Policy support and regulatory frameworks play a significant role in promoting waste management and recycling in agriculture. Governments can facilitate the adoption of sustainable practices by providing incentives, funding research, and establishing standards for waste treatment and recycling. By creating a supportive environment, policymakers can help farmers transition to more sustainable and efficient practices, ensuring long-term agricultural productivity and environmental stewardship.

Consumer awareness and demand for sustainably produced food can drive improvements in waste management practices. As consumers become more conscious of the environmental impact of their food choices, they increasingly seek products that align with their values, such as those with minimal packaging or certified sustainable practices. By meeting these expectations, farmers can enhance their market competitiveness and build trust with consumers, contributing to a more sustainable food system.

Waste management and recycling in agriculture offer significant opportunities for innovation and sustainability. By transforming waste into resources, reducing environmental impact, and conserving natural resources, farmers can contribute to a more sustainable and resilient agricultural landscape. Through collaboration, education, and a commitment to continuous improvement, the agricultural community can pave the way for a more sustainable and resource-efficient future.

Reducing Greenhouse Gas Emissions

The quest to reduce greenhouse gas emissions in agriculture is a critical endeavor in the fight against climate change. Agriculture is both a contributor to and a victim of climate change, making it imperative to develop strategies that minimize emissions while maintaining productivity. By addressing the sources of emissions, adopting innovative practices, and fostering a culture of sustainability, the agricultural community can play a pivotal role in mitigating climate impacts.

Agricultural emissions primarily originate from three sources: enteric fermentation in ruminants, manure management, and the use of synthetic fertilizers. Enteric fermentation, a natural digestive process in cattle and other ruminants, produces methane, a potent greenhouse gas. To tackle this, farmers can explore dietary modifications, such as adding fats or oils to feed, which can suppress methane production. Research into feed additives, such as seaweed, has shown promising results in reducing methane emissions significantly. By optimizing feed efficiency and improving herd management, farmers can lower the overall emissions from livestock operations.

Manure management is another area where emissions can be reduced. The decomposition of manure releases methane and nitrous oxide, both of which contribute to climate change. Implementing anaerobic digestion systems can capture methane for energy production, transforming waste into a valuable resource. Additionally, composting manure under controlled conditions can minimize emissions and produce nutrient-rich compost for soil enhancement. Proper storage and application techniques are essential to prevent emissions and nutrient runoff, ensuring that manure is managed sustainably.

The use of synthetic fertilizers in crop production is a significant source of nitrous oxide emissions. To reduce reliance on these fertilizers, farmers can adopt practices such as precision agriculture, which involves using technology to apply the right amount of nutrients at the right time. This approach minimizes waste and ensures that crops receive the nutrients they need without contributing to excess emissions. Integrating cover crops and crop rotations can enhance soil fertility naturally, reducing the need for chemical inputs. Leguminous cover crops, for example, can fix atmospheric nitrogen into the soil, providing a natural source of this essential nutrient.

Conservation tillage is another effective strategy for reducing greenhouse gas emissions. Traditional plowing releases stored carbon from the soil into the atmosphere, contributing to emissions. By adopting conservation tillage or no-till practices, farmers can preserve soil carbon and improve soil health. These practices also enhance water retention, reduce erosion, and promote biodiversity, contributing to a more resilient farming system.

Agroforestry, the integration of trees and shrubs into agricultural landscapes, offers multiple benefits for emission reduction. Trees act as carbon sinks, absorbing carbon dioxide from the atmosphere and storing it in their biomass. By incorporating agroforestry practices, farmers can sequester carbon, enhance biodiversity, and provide additional income streams through products like timber, fruits, and nuts. Selecting tree species that complement existing crops and livestock can optimize the benefits of agroforestry systems.

Water management also plays a role in emission reduction. Efficient irrigation practices, such as drip or subsurface irrigation, reduce water waste and the energy required for water pumping. Additionally, improving water management can enhance soil health, leading to increased carbon sequestration. Farmers can invest in water-saving technologies and practices to conserve resources and minimize their environmental impact.

Education and collaboration are vital for the successful implementation of emission-reducing practices. Farmers need access to resources, training, and support to adopt new technologies and strategies. Extension services, research institutions, and peer networks can provide valuable guidance and foster a culture of innovation and sustainability. By sharing knowledge and experiences, the agricultural community can collectively advance toward more sustainable practices.

Policy support and incentives are crucial for encouraging emission reduction in agriculture. Governments can facilitate the transition to sustainable practices by providing funding, developing standards, and supporting research into innovative technologies. By creating a supportive regulatory environment, policymakers can help farmers reduce emissions while maintaining productivity and profitability.

Consumer awareness and demand for sustainably produced food can drive improvements in emission reduction practices. As consumers become more conscious of the environmental impact of their food choices, they increasingly seek products that align with their values, such as those with eco-certifications or carbon-neutral labels. By meeting these expectations, farmers can enhance their market competitiveness and build trust with consumers, contributing to a more sustainable food system.